FLAT-WORLD
FICTION

FLAT-WORLD

LILIANA M. NAYDAN

THE UNIVERSITY OF GEORGIA PRESS | ATHENS

FICTION

Digital Humanity in Early
Twenty-First-Century America

© 2021 by the University of Georgia Press
Athens, Georgia 30602
www.ugapress.org
All rights reserved
Designed by Kaelin Chappell Broaddus
Set in 10.5/13.5 Garamond Premier Pro Regular by Kaelin Chappell Broaddus

Most University of Georgia Press titles are
available from popular e-book vendors.

Printed digitally

Library of Congress Cataloging-in-Publication Data

Names: Naydan, Liliana M., author.
Title: Flat-world fiction : digital humanity in early twenty-first-century America / Liliana M. Naydan.
Description: Athens : The University of Georgia Press, [2021] | Includes bibliographical references and index.
Identifiers: LCCN 2021021553 | ISBN 9780820360553 (hardback) | ISBN 9780820360560 (paperback) |
    ISBN 9780820360577 (ebook)
Subjects: LCSH: American fiction—21st century—History and criticism. | Technology in literature. | Literature
    and technology—United States—History—21st century. | Literature and society—United States—History—
    21st century. | Digital media—Social aspects. | LCGFT: Literary criticism.
Classification: LCC PS374.T434 N39 2021 | DDC 813/.609356—dc23
LC record available at https://lccn.loc.gov/2021021553

For Nina Naydan-McAsey
See. Touch. Hold.

CONTENTS

ACKNOWLEDGMENTS ix

Introduction. American Literature and Digital Technology in the New Millennium 1

Chapter 1. Relationships with Technology in Gary Shteyngart's *Super Sad True Love Story* and Kristen Roupenian's "Cat Person" 19

Chapter 2. Searching History in Thomas Pynchon's *Bleeding Edge* and Jennifer Egan's *A Visit from the Goon Squad* 46

Chapter 3. The Digital Divine in Joshua Ferris's *To Rise Again at a Decent Hour* and Jonathan Safran Foer's *Here I Am* 76

Chapter 4. Cybercapitalism in Don DeLillo's *Cosmopolis* and Dave Eggers's *The Circle* 111

Chapter 5. National Divides and Digitization in Zadie Smith's "Meet the President!" and Mohsin Hamid's *Exit West* 146

Conclusion. Flat-World Fiction and the Textured Future 179

NOTES 189

WORKS CITED 197

INDEX 211

ACKNOWLEDGMENTS

This book would not have been possible without the support of my family, friends, and colleagues. In particular, I want to thank my mother, Roxanne, who gave me insight into the work of Mark Rothko and offered me everyday support through conversation; my father, Michael, who talked me through the nuances of Russian literature and read my manuscript drafts many times, giving me suggestions about sentences and ideas; my husband, Jim, who listened to my ideas about digitization and talked me through challenges with writing; and my brilliant and bold daughter, Nina, who inspired this project with her dual enchantment with my smartphone and print books.

In addition, I'm grateful for the support from past and current administrators at Penn State, including Andy August, Friederike Baer, Damian Fernandez, and Roy Robson (who helped me with my introduction). They believed in my project and supported it with thoughtful counsel, a Summer Faculty Fellowship, an Outstanding Research Fellowship, and a Faculty Development Grant.

I'm also thankful for help I received from my colleagues and friends at and beyond Penn State, among them Dana Bauer, Ralph Clare, Mike Kagan, Charity Ketz, Mary Naydan, Carroll Pursell, and Matt Rigilano. Their insights helped me familiarize myself with everything from digital humanities to Einstein's theories.

Finally, thanks to Greg Clingham for help with my book proposal; to Walter Biggins for seeing potential in that proposal; to Beth Snead, Jon Davies, and others at the University of Georgia Press for guiding me through to publication; to Daniel Simon for his excellent copyediting of my book manuscript; to David Prout for indexing my book; and to Stacey Olster for continuing to graciously answer my old-fashioned emails and calls to her landline many years after my time at Stony Brook. She may lack social-media connections, but she has never failed to help me make useful connections in my scholarship.

An earlier version of part of chapter 5 of this book appeared as "Digital Screens and National Divides in Mohsin Hamid's *Exit West*," *Studies in the Novel*, vol. 51, no. 3, 2019, pp. 433–451. Copyright © 2019 by Johns Hopkins University Press and the University of North Texas.

FLAT-WORLD FICTION

INTRODUCTION

American Literature and Digital Technology in the New Millennium

At the dawn of the third millennium, in the aftermath of what print-media mogul Henry Luce deemed the American Century,[1] digital-age technological developments changed America and Americans in profound ways. And, in turn, they influenced American literature, which found itself implicated in the developing digital world, threatened by it, and also attempting to represent and critique it. As Carroll Pursell suggests in *Technology in Postwar America*, over the course of the twentieth century, technology enabled the United States to attain global prominence, particularly with the invention and detonation of the atomic bomb. As the twentieth century progressed, "the United States devised and supported a regime of technology that both convinced and compelled its own citizens, as well as distant peoples, to adjust to a globalized economy, culture, and political order designed to be very much in the American nation's favor" (Pursell ix). But as the twentieth century turned to the twenty-first, technology that Americans had celebrated came to undercut notions of America—and of nations in general—and the distinct sense of Americanness that those notions had helped create. Americans and global citizens alike saw the multifarious results of the digital revolution that Silicon Valley fueled. As they grew comfortable with networked personal computers, learned to ask questions of Jeeves and Google, and bought cell phones that would soon altogether change what the word *phone* means, they watched news stories about Y2K, the dot-com bust, digital identity theft, cyberbullying, and cyberterrorism. And they saw the emergence of what Thomas L. Friedman calls in *The World Is Flat: A Brief History of the Twenty-First Century* "Globalization 3.0," a new

| 1

era in global history that is marked by digital developments (10). In other words, Americans saw that the array of flat digital screens that were coming to surround them had the capacity not only to reach around the world but to reach deep into human consciousness.

This book addresses representations of the digital revolution and the social and ethical concerns that it created and continues to create in mainstream literary American fiction and fiction written about the United States in the first two decades of the twenty-first century, from the time of the Y2K crisis to the onset of the Covid-19 crisis, which forced many citizens of the globe to go online. I argue that Don DeLillo, Jennifer Egan, Dave Eggers, Joshua Ferris, Jonathan Safran Foer, Mohsin Hamid, Thomas Pynchon, Kristen Roupenian, Gary Shteyngart, and Zadie Smith use their literary texts to contemplate the problems and possibilities of digital devices and media that critics say are threatening to eradicate old-media print culture—a culture in which they continue to participate.[2] They explore how human relationships with digital devices and media transform identities as well as human relationships with one another, history, divinity, capitalism, and nationality. And they create accessible literary road maps to our digital future that complement and expand on work by historians, philosophers, and social scientists. Although these authors vary in the degree to which they posit that Americans or global citizens ought to embrace digitization and its social and cultural byproducts or remain skeptical of them, their views add depth to the conversation about the emergent flat world that Friedman idealizes. These writers show through fiction that technology is political. Further, they position twenty-first-century literature written in a purportedly post-text historical moment as having noteworthy importance because it can aestheticize unwebbed and more genuinely connected alternatives to the webbed and disconnected present. It can invite readers to develop philosophies of technology that acknowledge life's texture. It can also invite them to understand and embrace human hybridity and work to realize a more socially just and responsible future.

American Progress and the Politics of the Digital Age

Is the world really flat? Just four years into the twenty-first century in a so-called brief history that begged to be written because dramatic changes had redefined American life, Friedman suggested that it is. The first edition of *The World Is Flat*, published in 2005, greeted its readers with a cover that portrays two massive, Columbus-era ships that teeter at the edge of a waterfall, apparently poised to sail off the watery end of the earth. In the book, Friedman argues that a metaphorical flatness develops around the time that the World Trade Center falls flat in lower

Manhattan due to the 11 September 2001 terrorist attacks. As Friedman explains, discussing the United States' place in the current phase of globalization, which he terms Globalization 3.0, "The global competitive playing field [is] being leveled. The world [is] being flattened" (8). According to Friedman's theory of history, it is largely flattening due to digitization, a postanalog form of computer technology that relies on the translation of information into ones and zeros. It is flattening as a result of the propagation of largely American-born digital devices and media, for instance Netscape and Wikipedia, each of which, in Friedman's book, takes on a transcendent quality that speaks to David F. Noble's characterization of technological enchantment as "rooted in religious myths and ancient imaginings" (3). Coming into conversation with like-minded new media enthusiasts, Friedman contends that "Globalization 3.0 makes it possible for so many more people to plug in and play, and you are going to see every color of the human rainbow take part" (11).[3] It allows for what Friedman characterizes as the "empowerment of individuals to act globally" and attain what he calls a globalized version of the American Dream (11).

Flat earthers, too, think the world is flat, though in a literal way that complements Friedman's metaphor. Many of them are members of modern flat earth societies, and they believe that the earth has the shape of a pancake and not a sphere. They believe it has an icy circumference, not polar icecaps. And a recent study by Ashley Landrum of Texas Tech and a group of coresearchers finds that the number of flat earthers is on the rise in large part thanks to digital media. According to Ian Sample, a journalist who reports on Landrum et al.'s conference paper in a 2019 *Guardian* article, interviews with thirty attendees at a 2017 conference for flat earthers in North Carolina revealed that most of them "had been watching [YouTube] videos about other conspiracies, with alternative takes on 9/11, the Sandy Hook school shooting and whether NASA really went to the moon, when YouTube offered up Flat Earth videos for them to watch next."[4] According to Sample's interview with Landrum, one video in particular, American writer and producer Eric Dubay's "200 Proofs Earth Is Not a Spinning Ball," seems to be most effective in persuading them of the earth's flatness. The video "offers arguments that appeal to so many mindsets, from biblical literalists and conspiracy theorists to those of a more scientific bent" (Sample).

Even incredulous humanities scholars are increasingly embracing notions that the world—or at least the future of academia—is flat. Through their work, they have given birth to the digital humanities, a new area of scholarly humanities work that complements academic ventures in the virtual world such as Massive Open Online Courses (MOOCs) and that gets its name from William Pannapacker's "The MLA and the Digital Humanities," a 2009 blog post written at that year's Modern Language Association convention in Philadelphia and published online by the *Chronicle of Higher Education*. Reflecting on the grim state of the discipline of English following the 2008 financial crisis, Pannapacker suggests in his post that "our profession needs revitalization." To that end, he views digital humanists as poised to revitalize literary studies through their work, which involves the remediation of print texts into flat, digital form and the utilization of digital methods to study literary elements of these digitized texts. As Susan Schreibman, Ray Siemens, and John Unsworth add in their preface to *A New Companion to Digital Humanities*, the area of humanities work, which some scholars view as a field, is also broadening. It now, too, includes "the cultural study of digital technologies, their creative possibilities, and their social impact" (Schreibman et al. xvii). Hence the area is poised, as some scholars see it, to rehabilitate the dying disciplines that fall under the umbrella of humanities. Through digitally flattening primary and secondary sources, digital humanists hope to texturize literary studies and other fields that the MLA and similar professional organizations support. They hope to counter what Pannapacker views as "the doom and gloom" of English, a field that is in steady decline in the United States.

These visions of the flat world and future that Friedman, flat earthers, and digital humanists present illuminate key paradoxes of our increasingly technological times. And Pursell spotlights the most noteworthy of these paradoxes when he observes that in the postwar period, there arose "an increased belief" among Americans that technology "held the key to a stronger, richer, healthier, and happier America" (xii). Technological progress may well parallel or constitute American progress, as the vast commercialization and domestication of new technologies, including televisions and personal computers in the late twentieth century, suggest. It may result in better living, which DuPont points to through its quintessentially American and well-known ad campaign, "Better Things for Better Living... Through Chemistry," later shortened to "Better Living Through Chemistry." But, paradoxically, it may not, as even the idealistic Friedman acknowledges in observing that there "are two ways to flatten the world. One is to use your imagination to bring everyone up to the same level, and the other is to use your imagination to bring everyone down to the same level" (447). As Friedman's critics see it, *The World Is Flat* reads as a justification of American outsourcing, American offshoring, and corporate American modes of labor exploitation that only masquerade as socially just acts of leveling the playing field and making American life better. They see his neoliberal book as standing in stark contrast to more progressive ideas such as those that Ian Bremmer presents in *Us vs. Them: The Failure of Globalism*, which argues that "ongoing political, economic, and technological changes around the world" are "widening divisions" between the privileged few and the disenfranchised many (6).

Moreover, the rise of flat earthers due to YouTube highlights a second noteworthy paradox of the times: the notion that an increase in access to information produces a more educated or intelligent society or a better United States when it paradoxically may not. This paradox manifests in part because information that the internet provides requires a sophisticated kind of information literacy that digital citizens may lack, a point that numerous contemporary critics of digitization articulate. For instance, in *The Internet of Us: Knowing More and Understanding Less in the Age of Big Data*, Michael Patrick Lynch argues that "information technologies, for all their amazing uses, are obscuring a simple yet crucial fact: greater knowledge doesn't always bring with it greater understanding" (6). And James Bridle makes a similar argument in *New Dark Age: Technology and the End of the Future*. In Bridle's words, "we find ourselves connected to vast repositories of knowledge, and yet we have not learned to think. In fact, the opposite is true: that which was intended to enlighten the world in practice darkens it" (10). As Bridle continues, "The abundance of information and the plurality of worldviews now accessible to us through the Internet are not producing a coherent consensus reality, but one riven by fundamentalist insistence on simplistic narratives, conspiracy theories, and post-factual politics" (10–11). They are producing flat earthers in two senses of the term: first, digital natives who experience life predominantly through their flat screens, and second, misguided citizens who believe that baseless online propaganda is true.

Furthermore, digital humanists' efforts present a third paradox. Their seemingly cutting-edge work may result in progress for humanities fields, but, paradoxically, it may not. Indeed, many digital humanists face staunch criticism: accusations that they are flattening out the textured, analytic work of humanities disciplines. They face criticism that emerges despite digital humanists' efforts to add depth to what it means to work as a scholar in the humanities. As Daniel Allington, Sarah Brouillette, and David Golumbia suggest in "Neoliberal Tools (and Archives): A Political History of Digital Humanities," a much-discussed 2016 article in the *Los Angeles Review of Books*, digital humanities discourse, like "the rhetoric surrounding Silicon Valley today," characterizes "technological innovation as an end in itself and equates the development of disruptive business models with political progress." And this area of humanities research facilitates "the neoliberal takeover of the university" (Allington et al.). It plays what Allington, Brouillette, and Golumbia call "a leading role in the corporatist restructuring of the humanities." More recently and less contemptuously, Nan Z. Da agrees. In "The Digital Humanities Debacle," published in the *Chronicle of Higher Education* in 2019, she observes that the branch of the digital humanities that applies quantitative methods to literary studies has "generated bad literary criticism" and "tends to lack quantitative rigor." As

Da continues, "Its findings are either banal" or "not statistically robust." In other words, these scholars' findings fall flat and their research feels flat, at least in Da's view, because they "treat literary data in vastly reductive ways, ignoring everything we know about interpretation, culture, and history."

Paradoxes such as these emerge in the so-called flat world precisely because of its flattening, and they in part function to illuminate the always already political nature of digital technology—the notion that digital devices and media have developed and been used within a complex sociopolitical context since the politically contentious Cold War–era American moment of their birth. As a result of the arms race between the Soviet Union and the United States, the modern computer was born,[5] as was the internet in the form of ARPANET, America's first major computing network. Funded by what Fritz in Thomas Pynchon's *Inherent Vice* calls "government money" and designed by the U.S. Department of Defense's Advanced Research Projects Agency as a defense weapon, ARPANET was the product of paranoia of the kind that powers Pynchon's literary imagination (54). And it was also the product of white male American scientists, a point that the narrator of the second chapter of Colum McCann's *Let the Great World Spin* makes. In the narrator's words, "White men, all," from elite institutions such as Stanford and MIT, were "developing the dream of the ARPANET," which came to allow for communication in case of a Soviet nuclear attack (McCann 83). It, too, allowed for the exchange of scientific information by researchers—even though this militaristic history is all but invisible to contemporary Americans who predominantly surf the commercial web to work, shop, or play.

Recent history also sheds light on the notably political nature of digitization. For instance, whistleblowers such as Chelsea Manning who post to WikiLeaks, founded by the highly controversial Julian Assange, function as digital-age watchdogs amid the decline of traditional journalism, which is in decline because of digitization. Like other WikiLeaks posts, Manning's post of a video of U.S. soldiers shooting at and killing Iraqi civilians from a helicopter instigated public dialogue about reprehensible militaristic and governmental activity. Along the same lines, participants in online social movements such as the #MeToo movement, which began through the important activist work of Tarana Burke, have opportunities to produce transformative results as well. In late 2017, at the digital behest of television actress Alyssa Milano, tens of thousands of women exposed acts of sexual harassment and assault by famous men, among them American authors Junot Díaz and Sherman Alexie. And women inevitably did so in the face of critics who see online activism as reinforcing or exacerbating problematic existing power dynamics or who characterize online tweets and posts about social concerns as slacktivism—a brand of activism that exists for slackers and a pale imitation of social movements of old such as the civil rights movement, which brought activists out into U.S. streets.[6] As Antonia Malchik explains in a 2019 *Atlantic* article, "The Problem with Social-Media Protests," "The civil-rights movement took a decade to get to the March on Washington—time that Martin Luther King Jr. and his colleagues spent forming and deepening social connections, strengthening and testing the fiber of their movement. By contrast, mass protests such as Occupy Wall Street formed rapidly but then, lacking that underlying resilience created over time, often lost focus, direction, and, most important, their potential to effect change." As Malchik concludes, "New eras of protest will have to learn how to combine the ease and speed of online connectivity with the long-term face-to-face organizing that gives physical protest its strength and staying power."

Less visibly but still notably, the politically charged nature of digitization is both shaped by and gives shape to theories and philosophies of technology that emerge alongside digital developments. Since the publication of Martin Heidegger's "The Question Concerning Technology," which suggests that the ideal "relationship" with technology is "free," or one in which humanity is not enslaved by technology, questions of power dynamics and privilege have been part and parcel of phenomenological thought about technological devices (305). In contemporary American literary terms, to quote DeLillo's words from *Underworld*, humans in relationships with technology may feel in "the grip of systems" (825). And building on Heidegger's foundational work, American philosophers such as Don Ihde explore the political underpinnings and implications of human relationships with technology. As Ihde puts it in *Technology and the Lifeworld: From Garden to Earth*, technology can become embodied, or it can connect with a human to mediate the world, for instance as eyeglasses connect or merge with our eyes so we can see the world as it exists (73). It can connect with our thinking, or become hermeneutic, by merging with the world to make the world readable to humans, for instance as a thermometer merges with temperature so we can read and interpret that temperature (85). Or it can emerge as an alterity, as an Other to humans in the postcolonial sense (97).

Perhaps most notably among philosophers and theorists, Donna Haraway in "A Cyborg Manifesto: Science, Technology, and Socialist Feminism in the Late Twentieth Century" politicizes digital technology in Cold War–era American scholarly thought. She argues that nonimplanted, hybrid humans who are disenfranchised by capitalism should be celebrated for the mixtures they embody, for the socialist and feminist values they propagate, and for the potential they have to generate a countercapitalist world of unity and inclusion. In other words, Haraway suggests that the state of being a cyborg is metaphorical. And in doing so, she sets the stage for Andy Clark's argument that humans are "natural-born cyborgs"—hybrid beings who lack technological implants but who are "forever ready to merge" their "mental activities with the operations of pen, paper, and electronics" (*Natural-Born Cyborgs* 6).

Moreover, Haraway suggests that the metaphorical cyborgs she theorizes should be celebrated for their inherently political hybridity, which can be understood in two key ways. First, hybridity is the sort of postcolonial feature of amalgamated and ever-liminal human identity and culture that Homi K. Bhabha memorably celebrates in *The Location of Culture*, a text that uses the metaphor of a stairwell to characterize hybridity as "a liminal space" defined by "temporal movement" that "entertains difference without an assumed or imposed hierarchy" (5). As Bhabha elaborates, to be hybrid is to be "neither the one thing nor the other" but an amalgam (49). It is an increasingly debated phenomenon that, according to Peter Burke, some critics critique for its eradication of "regional traditions" and "local roots" (7). And other critics interrogate it, among them Anjali Prabhu, who, in *Hybridity: Limits, Transformations, Prospects*, tests the claim that hybridity provides "a way out of binary thinking" and is unequivocally liberatory (1). Second, hybridity is an unsettling byproduct of digital times in which humans live hybrid lives by moving between virtual and material realities, as Ihde notes. In *Bodies in Technology*, Ihde explains that the "older Cartesian worry" over "whether or not we could be deceived by a cleverly contrived robot, a look-alike," gives way to new anxieties over "whether hyperreality is such that 'reality isn't enough anymore'" (12).

Notions of digitization as a political phenomenon inform conversations about American relationships with digital technology and media as distinctively new and anxiety-inducing threats to American life.[7] For instance, American philosopher Hubert L. Dreyfus warns against overreliance on the internet in "Anonymity versus Commitment: The Dangers of Education on the Internet." Using Søren Kierkegaard's condemnation of the press in *The Present Age: On the Death of Rebellion* as a springboard for his own critique, Dreyfus hypothesizes that late twentieth-century humanity should "resist the nihilistic pull of the new network culture" because the World Wide Web can never foster meaningful education for its users (646). And American digital-media pioneer turned philosopher Jaron Lanier makes a similarly critical point about the web in *You Are Not a Gadget: A Manifesto*. Echoing Heidegger's seminal consideration of human relationships with technology, he notes that the problem with the digital revolution, which brought about the web's existence, is that it enslaves people while liberating devices and media. In Lanier's words, "Something started to go wrong with the digital revolution around the turn of the twenty-first century. The World Wide Web was flooded by a torrent of petty designs sometimes called web 2.0. This ideology promotes radical freedom on the surface of the web, but that freedom, ironically, is more for machines than people" (3).

Political conceptions of digitization likewise inform scholarly conversations about human relationships with digital devices and media in social science disciplines. For instance, in *Pressed for Time: The Acceleration of Life in Digital Capitalism*, a book that evokes Henry Adams's sense of changed time at the start of the twentieth century,[8] Judy Wajcman studies perceptions of the faster pace of life in the digital age and interrogates why we develop relationships with "digital devices to alleviate time pressure and yet blame them for driving it" (2). She argues that in contemporary capitalist culture, being "busy is valorized, while having too much time on one's hands signifies failure" (170). As a result, humans and not digital devices are responsible for creating "overloaded" states, which present a particular problem for women who engage in "unpaid work" more often than men (169, 168). More notably, MIT professor of science and technology Sherry Turkle examines ways in which human relationships with digital devices and media are changing humanity in her book series on computers and people. In *The Second Self: Computers and the Human Spirit*, the first work in the series, Turkle argues that computers are changing human nature, and perhaps not in ways that humans want to be changed. She presents findings from interviews she conducted with computer users ranging from children to engineers to gain insight into the "kind of people" humans are becoming because of "the computer as partner in a great diversity of relationships" (19, 20). Drawing attention to the connection between the politics of digitization and identity politics, she argues that the "computer has become an 'object-to-think-with'" that is "entering into our thinking about ourselves" (27, 29). Furthermore, in *Alone Together: Why We Expect More from Technology and Less from Each Other*, Turkle argues that relationships with digital devices and media are influencing humans' relationships with one another by allowing us to "hide from each other, even as we are tethered to each other" (1). As Turkle elaborates, as "we instant-message, e-mail, text, and Twitter, technology redraws the boundaries between intimacy and solitude," and as we "build a following on Facebook or MySpace," we "wonder to what degree our followers are friends" (11). Online connections come to substitute for both solitude and community in real life, creating "a new state of the self" as opposed to just a second self (16). Thus, digital technology may flatten people "into personae" in detrimental ways (18), rendering humanity as being in need of meaningful connections, which Turkle considers in *Reclaiming Conversation: The Power of Talk in a Digital Age*.

The inherently political relationships with digital devices and media that these philosophers and social scientists discuss produce what numerous authors of literature and literary critics see as a disintegrating American nation and a flat world that is flat because it is boring, superficial, corporate, and increasingly devoid of humanity. As Robert W. McChesney argues in *Digital Disconnect: How Capitalism*

Is Turning the Internet against Democracy, a link between corporate capitalism and digitization has come to exist even though the two need not be "bound together," and as a result, everything that the internet has "colonized" has also been colonized by capitalism (20, 3). As Zadie Smith intimates in "Generation Why?", her 2010 review of David Fincher's *The Social Network*, a film about the elite American origins of Facebook, this flat world is comprised of flattened-out people: onscreen, circumscribed, social-media-based shadows of textured reality. In Smith's words, "When a human being becomes a set of data on a website like Facebook, he or she is reduced. Everything shrinks. Individual character. Friendships. Language. Sensibility. In a way it's a transcendent experience: we lose our bodies, our messy feelings, our desires, our fears" ("Generation Why?").

The flat world is thus defined by a peculiar kind of depthlessness, which, to reference Jeffrey T. Nealon's discussion of post-postmodernism as involving intensification,[9] intensifies Fredric Jameson's description of the term in *Postmodernism, or, the Cultural Logic of Late Capitalism*. For Jameson, "the emergence of a new kind of flatness or depthlessness" is the emergence of "a new kind of superficiality in the most literal sense" (9). It involves an equalization of all things at the hands of capitalism, and it entails an attention to the surface rather than to modernist depth, which Jameson implicitly identifies as existing in modern art and literature. For citizens of the twenty-first century—for digital-age humanity, or (more simply) digital humanity—the culture of depthlessness that Jameson theorizes interweaves with the ubiquitous presence of flat digital screens that acquire the paradoxical capacity to reach deep into human consciousness in order to change human nature outright or to solicit resistance to the kind of change that they promise. In other words, depthlessness comes to connect in deep ways with aspects of the twenty-first-century American psyche. It challenges notions of human progress through technological development, and it sets the terms for ubiquitous human disconnection, which masquerades as connection to unsettling personal and political ends.

Flat-World Fiction as the New Science Fiction

Historically, science fiction as a predominantly American genre has taken on the work of philosophizing and critiquing scientific and technological developments that get conflated with or flattened into narratives of social progress.[10] According to David Seed, there came to exist a strong association between science fiction and "the evolution of technology, by which is usually meant tools or implements" (47). And Cold War–era science fiction focused particularly on developments in and anxieties involving digital computer technology, for instance in works such as Kurt Vonnegut's *Player Piano*, Philip K. Dick's *Vulcan's Hammer*, and, perhaps most notably, Stanley Kubrick's *2001: A Space Odyssey*, a 1968 film that showcases a showdown between man and an anthropomorphized, malfunctioning, and perhaps outright inimical computer known as Hal 9000. It likewise focused on the diminishing difference between humanity and computer technology through portrayals of robots of the sort that appear in Isaac Asimov's *I, Robot* and cyborgs of the kind that Martin Caidin fictionalizes in *Cyborg*. By the final decade of the Cold War, William Gibson brought digitization to the forefront of science fiction's concerns with the publication of "Burning Chrome," the short story that coined the now-popular term *cyberspace* and set the stage for *Neuromancer*, his 1984 cyberpunk novel about the subject. After the end of the Cold War, an array of novels and movies followed in Gibson's footsteps to establish cyberpunk as an important subgenre of science fiction, for instance Neal Stephenson's *Snow Crash* and *The Matrix*, a 1999 movie that portrays a futuristic nightmare which speaks through metaphor and allegory to the concerns of the digitally overdependent moment of its inception and reception. In the film, computer programmer Thomas Anderson, known among hackers as Neo, learns that the life he has believed himself to be living is merely virtual reality (projected by a system of cables and pods), which conceals dystopian conditions that resulted from a war between intelligent machines and humans. In the original film and the sequels that followed it—*The Matrix Reloaded* and *The Matrix Revolutions*—Neo enters into the Matrix as a Christ-like hero who leads a rebellion that delivers humans out of the clutches of programs and back into a free existence in real life.

Yet around the time that the Cold War ended, American authors who shunned the label of science fiction increasingly began to broach the subject of digital technology. And in the late 1980s and 1990s, Americans saw the birth of electronic literature that came to complement in its form the digital subjects, themes, and motifs that manifest in science fiction. Defined by N. Katherine Hayles as literature that is born digital, "a first-generation digital object created on a computer and (usually) meant to be read on a computer," electronic literature contains numerous subgenres (*Electronic Literature* 3). For instance, it contains hypertext fiction, which is "characterized by linking structures"; interactive fiction, which has strong "game elements" that blur the line between video games and fiction; and flash poems, which are "characterized by sequential screens that typically progress with minimal or no user intervention," among numerous other and ever-developing subgenres (*Electronic Literature* 6, 8, 28). Michael Joyce's *afternoon, a story* and Shelley Jackson's *Patchwork Girl, or, A Modern Monster*, both works of "first-generation" hypertext fiction, were among the earliest existing works of electronic literature, and they thereby remain among the most important works in the genre (Hayles, *Elec-*

tronic Literature 7). Joyce's piece, created in Storyspace, speaks to the sense of uncertainty that the digital revolution created by way of allowing its reader to click through different paths to learn the story of a technical writer who sees wreckage from a car crash that he comes to suspect involved his ex-wife and his son. By contrast, themes that emerge in Jackson's work speak to the digital revolution's relationship to science fiction and the pervasive stereotype in America that the digital age produces toys for boys. She presents *Patchwork Girl*, likewise created in Storyspace, as parts of a female human body that text and image stitch together much as Victor Frankenstein stitches the creature together in Mary Shelley's *Frankenstein*, which scholars identify as one of the first works of science fiction in the Western canon.

In turn, mainstream American literary fiction increasingly comes to address the subject of technology. Notably, this shift in the focus of mainstream literary fiction parallels a shift in the increasingly digitized process of its creation. As Hayles explains in "The Future of Literature: Complex Surfaces of Electronic Texts and Print Books," "almost all contemporary literature is *already* digital" because "print literature consists of digital files through most of its existence" (180). And, as Hayles continues, "print and electronic textuality" thereby "deeply interpenetrate one another" ("The Future of Literature" 181). The increasingly digital life that mainstream literature lives influences the digital subjects that this literature addresses near the Cold War's end, for instance in a work such as John Updike's *Roger's Version*, and particularly following its conclusion. In the 1990s, movies such as *The Net* and novels such as Richard Powers's *Galatea 2.2* and Don DeLillo's *Underworld*, the latter of which concludes with the juxtaposition of an image of the World Wide Web with the "thick lived tenor of things" in the world, fueled American anxiety about digitization (*Underworld* 827). They came to function as a new, mainstream variation on or complement to science fiction. Seed testifies explicitly to this shift in the literary landscape, observing that "an important sign of the shift in the status of science fiction has been the increasing willingness of so-called mainstream authors to adopt its themes and practices" (124). And Peter Boxall acknowledges it as well in his seminal study of mainstream twenty-first-century literature, albeit more tacitly than Seed. In Boxall's words, twenty-first-century "time is bent and crafted by the computer, the mobile phone, the satellite [and] the internet; by electronic communication at the speed of light," all of which mark "a new phase of modernity" (4). And, as Boxall continues, "the exponential acceleration in the speed of computing has produced the experience of a virtual, unlimited extension of the self, as rapid information exchange allows us to occupy space and time in entirely new ways" (90). Indeed, by the twenty-first century, in Vincent Miller's words, "the online sphere is no longer a realm separate from the offline 'real world,' but fully integrated into offline life" (1). Along the same lines, Laura J. Gurak and Smiljana Antonijevic observe that we "have now reached a time when the phrase 'digital rhetoric' is redundant" because nothing is "beyond the realm of the digital" anymore (497).

Numerous mainstream authors in and beyond America come to write what I call flat-world fiction, or twenty-first-century fiction about the ever-evolving hybrid world and the politically charged relationships that humans have with digital devices and media that pervade this world. Works such as M. T. Anderson's young adult novel *Feed*; John Barth's *Coming Soon!!!*; and Matt Beaumont's *e: A Novel* and its sequels, *The e Before Christmas* and *e Squared*, serve as examples of this new kind of fiction, which spotlights ways in which digital devices and media are an organic and integral part of everyday twenty-first-century life. Other examples of this new kind of fiction include Joshua Cohen's *Four New Messages* and *Book of Numbers*; Don DeLillo's *Cosmopolis* and *The Silence*; and Jennifer Egan's *Look at Me*, *A Visit from the Goon Squad*, and "Black Box." Yet other examples of flat-world fiction include Dave Eggers's *A Hologram for the King* and *The Circle*; Joshua Ferris's *To Rise Again at a Decent Hour*; Jonathan Safran Foer's *Here I Am*; Mohsin Hamid's *Exit West*; Barbara Kingsolver's *Flight Behavior*; and Tao Lin's *Shoplifting from American Apparel*. Still other examples include Thomas Pynchon's *Bleeding Edge*; Kristen Roupenian's "Cat Person"; Gary Shteyngart's *Super Sad True Love Story*; and Zadie Smith's "Meet the President!" All these works aestheticize and politicize digital devices, digital media, and human relationships with them. And they put them on display for contemplation by their audiences much as DeLillo places Douglas Gordon's *24 Hour Psycho* on exhibit in New York City's Museum of Modern Art in the fictional world of *Point Omega*, a novella that speaks to flat-world fiction as I define it for its focus on screen culture. Just as the screen of Gordon's 1993 art installation appears in the frame tale of *Point Omega* while functioning as the novella's metaphorical centerpiece, so, too, do the digital devices and media of flat-world fiction—devices and media that conspicuously descend from the screen represented by DeLillo in *Point Omega*—occupy both peripheral and central places in contemporary American life. Because these devices and media are everywhere, there exists an illusion that they are nowhere and on the periphery, as the screen of DeLillo's novella is on the periphery in the frame tale. But many contemporary Americans would feel paralyzed without these quintessentially central devices and media.

Centering digital devices and media as DeLillo centers the screen in *Point Omega*, my study builds on John Johnston's *Information Multiplicity: American Fiction in the Age of Media Saturation* and Zara Dinnen's *The Digital Banal: New Media and American Literature and Culture*, literary analyses that consider the ways in which authors of mainstream literary texts address features of digital-age life. And this book also comes into conversation with works on the relationship between digital and print culture such as Jessica Pressman's *Digital Modernism: Making It New in New Media* and Adam Hammond's *Literature in the Digital Age: An Introduction*. In this book, I focus on mainstream literary fiction about human relationships with digital devices and media by American authors and authors writing about the United States as part of the digitally connected, globalized world. In doing so, I attend to the mounting tension involving national or specifically Americanist lenses for literary analysis in the globalized world as well as the largely American origins of digitization as a phenomenon that perpetuates globalization. Although this framework comes into conversation with international or global perspectives on digitization, it focuses on twenty-first-century American experiences and outside perspectives on those experiences. It thereby reveals the ways in which digitization has changed authorial and public feelings about the United States in the aftermath of the American century.

Specifically, I focus on the perspectives of Don DeLillo, Jennifer Egan, Dave Eggers, Joshua Ferris, Jonathan Safran Foer, Mohsin Hamid, Thomas Pynchon, Kristen Roupenian, Gary Shteyngart, and Zadie Smith. I consider the ways in which these diverse authors aestheticize and critique a widespread digital culture that threatens print culture and screens humans from one another. And I consider ways in which, in the process, they produce politically charged philosophies of digital technology in literary form that build on ideas about hybrid identity that Haraway and Bhabha put forth and that present alternatives to the idealistic vision of digitization that Friedman presents. These authors deconstruct the metaphorical binary code of the conversation about technology, exploring everyday negotiations with digitization and its influence on American thinking, particularly about love, history, religion, capitalism, and nationality, subjects that consistently arise in the novels and short stories that I address and subjects around which I thereby organize this book. Amid dialogue about digitization that Turkle sees as "steeped in the language of liberation and utopian possibility," I see these authors as showing a concern for humanity, community, social responsibility, and social justice in the American and global future, but in ways that complement the work of science fiction (*Life on the Screen* 246). Through their political texts, they put a premium on making social progress according to their unique definitions of it; they avoid conflating social progress with technological progress as Americans have historically conflated them; and they invite their readers to do the same in the real hybrid world, which, to cite Peter Burke, is hybrid as a result of extant fusions within (and, I would add, between) "artefacts, practices and finally people" (13).

In chapter 1, I consider ways in which Gary Shteyngart in *Super Sad True Love Story* and Kristen Roupenian in "Cat Person" comment on human relationships with technology as Heidegger does in "The Question Concerning Technology," suggesting that they hybridize and inhibit prospective human love relationships in detrimental ways. For Shteyngart, digital devices and media foster superficiality, immaturity, and illiteracy. And predominantly because of them, Shteyngart's fictionalized, near-future United States experiences political and financial collapse, as does the digitally mediated love relationship between Shteyngart's largely technophobic protagonist and his digitally connected Korean American love interest. Similarly, for Roupenian, digital devices and media screen out the complex reality of individual identity as a building block for a relationship, allowing users such as her privileged college-student protagonist and her protagonist's apparently less privileged and unattractive hookup to blur the distinction between reality and fantasy; misread and dehumanize one another through stereotypes and assumptions; and develop illiteracies of their own motivations and desires. Both Shteyngart and Roupenian indicate that literature can help remedy unsettling digital-age conditions. Whereas Shteyngart suggests that literacy and critical reading that exposes intertextual relationships can help foster literacy toward cultural, religious, ethnic, and national Others and perhaps more mature human relationships among individuals, Roupenian intimates that literary works might offer valuable literacies of gender politics, power, and rape culture to digital citizens who seek to attain more equitable and thereby more mature and meaningful love.

In chapter 2, I build on my discussion of the problems and possibilities of digitally mediated romantic relationships by considering ways in which digital devices and media have altered human relationships with history. Through the lens of Thomas Pynchon's *Bleeding Edge* and Jennifer Egan's *A Visit from the Goon Squad*, I explore the ubiquity of nostalgia for the predigital past and ways in which that past might inform the construction of a counterexceptionalist American future. For Pynchon, this nostalgia produces unsettling cycles of American history and paradoxically allows for the eradication of the potentially didactic and inherently political past. In other words, nostalgia prompts Americans to remake the problematic past in the present and future without learning about problems and missteps in history. For Egan, this nostalgia may function to commemorate or celebrate aesthetics of old in the aesthetically flat digital age, specifically with regard to predigital-age music. But the valorization of the past produces problems akin to those that Pynchon explores. And these problems emerge through indiscriminate celebration of the digital future and the apparent progress it brings. Both Pynchon and Egan cast their texts as metaphorical road maps for the digital-age moment. They function as aesthetic objects that allow readers to make connections about history and engage with a hybrid world that combines virtual and material reality in more productive ways, most notably through countercorporate action. Whereas Pynchon advocates for a future of feminist solidarity in the face of a state of disconnection that enables the digital age and the age of terror, Egan advocates for viewing the present moment as an opportunity that invites the metaphorical composition of a counternarrative for the United States in the wake of the problematic exceptionalism of the American Century.

Chapter 3 focuses on religion as a particular aspect of history as Pynchon and Egan explore it in their respective works. Through readings of Joshua Ferris's *To Rise Again at a Decent Hour* and Jonathan Safran Foer's *Here I Am*, I consider the ways in which religious history, texts, and ideas shape American engagements with and conceptions of digital devices and media. As Ferris suggests, paradoxes and problems involving faith and texts on which faiths are based as well as American impulses for faith contribute to producing and understanding the peculiar posttruth, posthuman moment—a moment defined by contradictions, among them the notion that digital media deserve deification despite the disconnected communities and flattened-out, fractured, and misinformed citizens they produce. Similarly, Foer dramatizes the domestic and international implications of living a fragmented life in the increasingly digital twenty-first century, showcasing ways in which modern-day Americans live other lives through digital and nondigital means and thereby exist in a constant state of paradox—as emotionally and psychologically absent yet physically present in their everyday experiences. Both Ferris and Foer avoid advocating for old world religious faith in the new hybrid world. Instead, they invite their readers to turn to key aspects of religious life to find meaning and connect with aspects of the disconnected digital sphere. Ferris celebrates professional and humanitarian work that socially concerned believers might perform. He advocates for work toward the social good within and beyond America's bounds. And Foer invites readers to wrestle with religious and literary texts in accord with the meaning of Israel as his protagonist wrestles with these texts to find a maturity that leads to familial responsibility and more integrated ways of living in the hybrid world.

Chapter 4 builds on the connection between the digital as divine that I establish in chapter 3 by showcasing ways in which capitalism takes on transcendent, otherworldly, and spiritual qualities as corporate America increasingly comes to rely on, produce, and sell digital devices and media. In this chapter, I analyze Don DeLillo's *Cosmopolis* and Dave Eggers's *The Circle*, novels that focus on problems with cybercapitalism and cybercapitalist interests, which, for both authors, are inhumane. In DeLillo's novel, cybercapitalism sets the terms for reality, hybridizing and flattening otherwise textured life and warping notions of time. DeLillo showcases these contemporary transformations through aesthetic features of his text. Similarly, in Ferris's novel, cybercapitalism spreads via a quasifundamentalist faith in all things financial that corporate elites hold, and it transforms mass American conceptions of identity, humanity, and what it means to live in the natural world. Both DeLillo and Eggers advocate for the development of countercorporate values that, for instance, resist the veneration of data as opposed to the development of knowledge and that resist willing participation in a culture of surveillance. And both authors see literary works such as the novels they write as subtly involved in the propagation of these countercorporate values. Whereas DeLillo encourages reflection on mystery, history, and personal agency to avoid mechanical reproduction of the past as a phenomenon that has power to manipulate individuals in accord with corporate efforts to own the masses, Eggers encourages critical as opposed to corporate thinking, particularly about the countercorporate allegories and metaphors of his novel. He seeks to facilitate the production of humanitarian efforts that resemble his own. Through critical thinking, readers of these texts might envision alternatives to the United States' capitalist future in the face of prognostications of American progress through corporate-driven technological development that benefits corporations alone.

In chapter 5, I build on chapter 4's discussion of capitalism by considering the national divides that American capitalists transcend through globalization. In this chapter, I broaden the definition of what counts as American literature in the age of globalization through a discussion of works that imagine the United States as interconnected with other nations and as in part responsible for global crises—works written by Zadie Smith and Mohsin Hamid, authors who live in the United States for part of every year but who have quintessentially hybrid and not American identities. In "Meet the President!" and *Exit West*, Smith and Hamid respectively suggest that digital device and media usage may result in the transcendence of national borders to produce an interconnected world, but it also results in the dehumanization of marginalized or screened-out individuals. Specifically, Smith showcases ways in which digital devices and media empower privileged global citizens who devalue local as opposed to global life to reify social class divides and enact emotional, psychological, and physical violence against disenfranchised, local, lower-class citizens. Similarly, Hamid suggests that although refugees may rely on digital devices and media to maintain a sense of connection as they migrate across nations, these devices and media take on weaponlike qualities. They fragment their lives and allow xenophobic insiders to alienate and oppress them. Whereas Smith suggests that engaging with spiritual reality as a complement to physical and virtual reality can create more ethical thinking and more meaningful modes of connection than globalization creates by connecting citizens across nations, Hamid suggests that hybrid works of art such as his novel can help twenty-first-century humanity cultivate visions for socially just and meaningful interconnection. And, he indicates, meaningful interconnection runs counter to the superficial types of connections that globalizers foster.

In the conclusion to this book, I outline ways in which I see flat-world fiction as effectively texturizing twenty-first-century American literature, humanity, and life in the digital age, particularly through portrayals of hybridity in the complementary digital and social senses of the term. This literature thereby subtly influences the recalibration of values in the United States and fosters social justice of a sort that eludes the purview of Friedman's *The World Is Flat*. Specifically, this literature puts a premium on fostering information literacy, a type of literacy that flat earthers lack that is essential to informed civic and social engagement. It likewise puts a premium on moving between superficial and deep modes of reading that the surface-reading debate outlines by fostering multiliteracies for the hybridized times that digital humanities scholars prioritize—literacies of always already multimodal alphabetic texts, imagistic texts, print texts, and digital texts. Finally and most importantly, it puts a premium on fostering multiliteracies of hybrid individuals with diverse and intersectional social identities. Thus, I posit that although numerous cultural critics prophesy the death of books or characterize the present as a posttext historical moment of apocalyptic proportions, literary fiction as an increasingly hybrid medium remains vital in both senses of the term in the United States. It remains alive and remains essential to twenty-first-century American life because it allows readers to imagine and better understand Others. In turn, it invites them to feel and act on countercorporate social concern in a paradoxically connected yet disconnected and increasingly hybrid world.

CHAPTER 1

Relationships with Technology in Gary Shteyngart's *Super Sad True Love Story* and Kristen Roupenian's "Cat Person"

In "The Question Concerning Technology," a key philosophical text on technology, Martin Heidegger sets out to examine the "relationship" that humans have with technology, viewing this relationship as "free" and positive if it "opens our human existence to the essence of technology" (305). This notion of humans as having a political relationship with technology informs notions of relationships in general as they expand over the course of the twenty-first century because digital devices and media have come to mediate human relationships in striking ways. As Thomas L. Friedman suggests in "Online and Scared," a 2017 *New York Times* article that builds on the argument of *The World Is Flat*, "Cyberspace is now where we do [...] more of our dating" and "more of our friendship-making and sustaining." And the friendships and romances that individuals develop intermesh with their nondigital interactions, giving new meaning to hybridity as thinkers such as Donna Haraway in "A Cyborg Manifesto" or Homi K. Bhabha in *The Location of Culture* have theorized it. These friendships and romances may diminish notions of what a friendship or romance is and does, as proposed by Sherry Turkle. According to Turkle in *Alone Together*, virtual spaces blur the line between "intimacy and solitude" (12). And as Turkle indicates in *Reclaiming Conversation*, studies of digital media show that "our new efficient quests for romance are tied up in behavior that discourages empathy and intimacy" (180). Perhaps surprisingly, our quests for romance, too, seem tied up in global capitalism, as evidenced by Turkle's discussion of Liam, a New York City graduate student who sees "his romantic life in terms of product placement" because he sees himself as "the product" that he is "direct marketing" by using Tinder, a popular dating app (*Reclaiming Conversation* 182).

This chapter explores the problems and possibilities of hybrid relationships that bridge virtual spaces and the physical world through readings of Gary Shteyngart's *Super Sad True Love Story* and Kristin Roupenian's "Cat Person." These fictional works suggest that digital-age romances struggle to develop into loving relationships due to the immaturity and inequity that digital media propagate among prospective lovers. In the first part of this chapter, I focus on Shteyngart, a Russian-Jewish writer whose interracial and intercultural marriage complicates his twenty-first-century American identity in an increasingly globalized world.[1] Perhaps as a result, Shteyngart has consistently shown interest in hybridity in Bhabha's sense of the term and in relationships that involve it, be they hybrid because of individuals' differing social identities or as a result of digital technology.[2] For instance, in *The Russian Debutante's Handbook*, Shteyngart showcases hybridity through what Adrian Wanner argues is a "self-ironic performance of identity" (675). From the standpoint that his assimilated identity affords him, he tells the story of an "unassimilated and culturally alienated" twenty-five-year-old Russian-American Jew, Vladimir Girshkin, who has relationships with two American partners whose identities differ from his own dramatically (Wanner 676). By contrast, in "O.K., Glass: Confessions of a Google Glass Explorer," a 2013 *New Yorker* essay, Shteyngart showcases hybridity of a digital variety. In chronicling his own experience of winning "a Twitter contest run by Google to pick the first batch of Glass Explorers" and subsequently wearing and using his Google Glass, Shteyngart describes his own merger with digital technology. He notes that his iPhone "became a frightening appendage to a life of already sizable anxiety" as a "reproving parent," a "needy lover," and a "sadistic life coach," reminding him that whatever he "was doing, there were more fascinating things to be done." He, too, draws attention to his own increasing affection for the Google Glass device he wears. As he explains, when his Glass purrs "*Gamburrrger*" as a Russian translation of hamburger at his request, it resembles the harsh but beloved sound of his "grandmother at the end of a long hot day." And it makes him "all of a sudden" feel "something for this technology."

In part emerging out of Wanner's discussion of Shteyngart as a bicultural hybrid author, my analysis of Shteyngart considers relationships among twenty-first-century citizens, digital media, and literature in *Super Sad True Love Story*, a work of near-future fiction presented in a hybrid, old and new media epistolary form through diary entries, letters, and social media messages and set in an American nation that is experiencing political and financial collapse. In the novel, Shteyngart depicts characters in the globalized, ever-digitizing world who find commonality in their cultural hybridity while clashing over their respective perspectives on digital media and print books. He tells the story of thirty-nine-year-old Lenny Abramov, an unattractive print-book-loving Russian American of lapsed Jewish faith who courts the twenty-four-year-old Eunice Park, a petite, attractive Korean American who worships her GlobalTeens social media account and her cutting-edge "äppärät," a smartphone-like digital device that "beams holographic data," streams "video at eye level," ranks its wearers and those around them according to "categories such as Fuckability," and is nearly ubiquitous in the world of Shteyngart's novel ("O.K., Glass"). As the relationship between these two characters develops, digital media inhibit their ability to connect in a mature and lasting way. They hybridize their everyday interactions and foster superficiality, immaturity, and illiteracy to detrimental ends, as evidenced by intertextual relationships with literary works that Shteyngart creates to complement the romantic relationship on which his novel centers. Indeed, Shteyngart positions his readers to see what illiterate characters in his fictionalized future in particular are incapable of understanding: that illiteracy and misreading as digital media help circulate them have the capacity to inhibit mature love and that reading can help facilitate it. However, Shteyngart posits that human relationships with digital media can and should change. He suggests that if humans learn to engage with them strategically as opposed to obsessively, digital media can hybridize them to enriching ends. Along the same lines, he shows that literature's relationship with digital media is inevitably changing, and through negotiating with digital media, authors have the opportunity to further hybridize literature and increase their own and literature's relevance to twenty-first-century life. Ultimately, hybridized literary texts such as Lenny's and Shteyngart's can stimulate the human imagination as they prompt readers to puzzle over cunning blends of lies and truth and lead them away from super-sad ends. They can help readers find maturity through introspective contemplation and perhaps even mature love with hybrid Others to themselves. Moreover, they can help them develop ways of thinking and being that might counter capitalist dehumanization and cruelty.

In the second part of this chapter, I focus on Roupenian, a queer author who was a virtual unknown working on an MFA in creative writing at the University of Michigan when the *New Yorker* published "Cat Person" on 4 December 2017. Roupenian says she felt surprised by the online attention to her story, which was inspired by "a small but nasty encounter" she had with a person she met online ("Kristen Roupenian on the Self-Deceptions"). In the story, she narrates in the close third person the short-lived and largely digital-media-based relationship between Margot, an attractive and presumably white twenty-year-old college student, and Robert, a thirty-four-year-old, non-college-affiliated resident of the unnamed Ann Arbor–like college town where the story is set. According to Joe

Fassler, the story "went viral as almost no other work of short fiction has managed to do" ("The Joyce Carol Oates Story"). And as Roupenian explains in a 2019 *New Yorker* article, her sudden internet fame surprised her because she was a "thirty-something late millennial who had tweeted a grand total of twelve times in her life" ("What It Felt Like"). She found her Twitter account inundated with "a bunch of notifications from strangers" and her email inbox flooded by messages from new fans, old acquaintances, and requests for interviews ("What It Felt Like"). She then learned that her story became the second-most-read online *New Yorker* piece of the year, "trailing only Ronan Farrow's initial bombshell report on Harvey Weinstein's alleged sexual misconduct" (Roupenian, "With *You Know You Want This*"). Perhaps because of its timeliness, the story, as Roupenian explains (referencing the hotly debated dress photograph of 2015), "threatened to become the blue-dress/white-dress moment of the #MeToo era"—an era defined by feminist efforts countering sexual harassment and violence that began with accusations against then-acclaimed movie producer Harvey Weinstein ("What It Felt Like"). Women largely identified with the fictional work, seeing it as "a little too real," to reference Olga Khazan's words ("A Viral Short Story"). By contrast, some men felt attacked by it, as evidenced by the birth of the Twitter account *Men React to Cat Person* and #MenCatPerson, the hashtag men used to air their frustrations with feminist readings of the story and with what they identified as the fictionalized Margot's fat-shaming of Robert. Indeed, many of the story's readers, both men and women, misread the work of fiction as a nonfictional narrative. In Roupenian's words, the story's "status as fiction had largely got lost" in part because of "the story's realism" and also because "the story was shared again and again" online, "moving it further and further from its original context" ("What It Felt Like").

Building on Fassler's characterization of Roupenian as mining "the territory [...] between fantasy and reality" in *You Know You Want This: "Cat Person" and Other Stories*, Roupenian's first book, this analysis considers "Cat Person" as a meditation on the effects that digital devices have on romantic relationships, particularly women's experiences with them, in what Friedman refers to as the flat world. I argue that in the heavily mediated twenty-first century, romance involves the interplay of nondigital and digital communication, be it online dating of the sort that Roupenian portrays in "The Boy in the Pool" and "Death Wish" or be it text messaging, which shapes "Cat Person" as a hybridized work that resembles Shteyngart's *Super Sad True Love Story*. As a result of the hybridization of Roupenian's text and the hybridization of reality for characters within it, the distinction between reality and fantasy in lovers' relationships appears as increasingly complex. As Roupenian portrays lovers, they consistently rely on fantasy-based assumptions and stereotypes of prospective partners because digital screens screen out the multifaceted reality of human identity. They render their prospective lovers as enigmatic, misread, and dehumanized. However, Roupenian also suggests that digital media inhibit individuals from fully understanding themselves and their own motivations and desires, particularly toxic desires that pervade Roupenian's short-story collection. They act based on social expectations instead of acting on their feelings. As a result, they perpetuate social problems related to gender, sexuality, and power. Thus, Roupenian renders in fiction a dystopian counternarrative to Haraway's "A Cyborg Manifesto," which implies that technology can (and should) exist as a potentially empowering political tool for disempowered individuals because merger can produce equity. She renders the collapse of meaningful communication and love in an age of purportedly ubiquitous connection. And like Shteyngart, who puts a premium on literacy as a solution to problems that digital media propagate, she intimates that literary works, be they online or in print, might offer valuable literacies to digital citizens who seek to change or at least attain a more meaningful understanding of the hybridized world.

Hybrid Relationships with Cultural Others, Digital Media, and Literature in Gary Shteyngart's *Super Sad True Love Story*

In *Super Sad True Love Story* (2010), the focus of this first section, Shteyngart suggests that globalization—which fosters migration, the development of digital technology, and apparent connectivity in different forms—produces hybrid individuals in the dystopian and apparently interconnected future that his novel depicts. Against the backdrop of a politically and problematically interconnected American nation run by a Bipartisan Party comprised of Democrats and Republicans who cooperate toward their shared neoliberal goals, he stages the evolution of a connection of a different kind: a love relationship between the Korean American and culturally hybridized Eunice Park and the equally hybrid Russian American Lenny Abramov. Just as familial migrations shape Eunice and Lenny as hybrid individuals, so, too, do their respective engagements or disengagements with digital media. Lenny, who loves material objects such as his furniture and mementos, reveres his "Wall of Books" above all else and uses digital devices, including the outdated "retro äppärät" he wears near the novel's start, sparingly (*Super Sad* 52, 38). Although technology inevitably hybridizes him because engagement with it is unavoidable in Shteyngart's fictionalized future, Lenny attempts to sustain a counterdigital life because books help him make sense of his experiences and give him the kind of "inner life" that Shteyngart describes in *Little Failure* (3). By contrast, Shteyngart portrays Eunice as a foil to Lenny and as a peculiar mismatch of a partner for him because she relies on digital devices. In Shteyngart's words, Eunice is a "sleek digital creature" who wears around her neck an equally sleek äppärät, a parodic (to cite Linda Hutcheon's term) and homophonic representation of the apparat as the administrative system of the Communist Party in that it administers her life and the lives of its digital citizens (*Super Sad* 153).[3] She fixates on her cutting-edge digital device much as numerous virtually subhuman characters in Shteyngart's text do when, for instance, they mumble "lowly into their äppäräti" instead of engaging with the real world (8). As Raymond Malewitz sees it, Eunice serves as a prime example of a human in a "posthuman culture that wants nothing more than to escape from materiality into the digital realm" (112). She also bears a striking resemblance to the technophiles of numerous other works of fiction about technology, for instance Sam Bloch in Jonathan Safran Foer's *Here I Am* or Bill Peek in Zadie Smith's "Meet the President!"

By creating an intertextual relationship between *Super Sad True Love Story* and Octavia Butler's "Bloodchild," a short story that Shteyngart identifies as particularly influential on his thinking in "O.K., Glass," Shteyngart suggests that digital citizens such as Eunice sustain quasiromantic relationships with their digital devices. As Shteyngart explains, more so than Isaac Asimov's *Science Fiction* magazine or William Gibson's *Neuromancer*, "Bloodchild" (as science fiction) "caught" his "imagination" because although many "reviewers thought of the story as an allegory of slavery (perhaps influenced by the fact that Butler was African-American)," Butler "denied the claim" and "wrote that she thought of 'Bloodchild' as 'a love story between two very different beings'" ("O.K., Glass"). Shteyngart fashions *Super Sad True Love Story* as a postmodern parody of Butler's love story about the insectlike Tlics, which offer humans a preserve on their planet and in return ask humans to host their eggs while developing feelings of affection for those humans. According to Shteyngart's remarks, although the relationship between these species "is unequal and often gruesome," they clearly "need each other to survive," much as humans and technology appear to need one another to survive in *Super Sad True Love Story* ("O.K., Glass"). As Shteyngart observes, when he thinks of "our relationship with technology," he "cannot help but think of human and Tlic, the latter's insect limbs wrapped around the former's warm-blooded trunk, about to hatch something new" ("O.K., Glass"). Hence, depending on how one interprets his metaphor, he suggests that technology feeds on humanity or that humans feed on technology, overuse it, and outright abuse it. And in the process, they grow accustomed to interacting with apparently unfeeling, nonhuman partners. They thereby come to treat humans as inhuman, as evidenced, for instance, by Lenny's fear of the overweight "guy who registered *nothing*" on his äppärät, a man whom, Malewitz notes, "garners no sympathy" in Shteyngart's posthuman society because his appearance makes him an "inhuman" human (Shteyngart, *Super Sad* 34; Malewitz 115, 115).

Shteyngart suggests that digital media with which digital citizens have relationships inhibit hybrid individuals from communicating and connecting with other humans romantically in consistently meaningful ways. In *Super Sad True Love Story*, digital media set the terms for real-life interactions, for instance when Lenny and Eunice first meet in Rome. As Lenny describes his memory of their meeting, they "locked eyes for a millisecond, but it was enough time" for him "to download a million bits of sympathy" much in the way that a computer downloads data (194). And their relationship develops through a hybridized and often laughable mix of in-person and online conversation. Before Eunice joins Lenny in New York and even after she comes to stay at his apartment, they communicate through GlobalTeens, a social media platform that is akin to Facebook. And they, too, *avoid* communicating through GlobalTeens and in person. They are often *alone together*, to echo the title of Turkle's third book about computers and people. In other words, to paraphrase Turkle's argument, their bodies appear in the same physical space together, but Eunice frequently turns to digital means—her äppärät—to leave it.

Likewise, they present fragments or altered versions of their whole, purportedly true selves through digital writing because they can carefully select and "edit" their digital words, to reference Turkle's conceptualization of this phenomenon in *Reclaiming Conversation* (23). Through digital devices that function much as paramours do in and of themselves, Eunice maintains largely secretive correspondence with other prospective love interests and opts against telling Lenny about her digital and emotional connections. For example, she corresponds through GlobalTeens with David, a veteran of America's fictionalized war with Venezuela who shows romantic interest in her. And she corresponds with Lenny's septuagenarian but deceptively young-looking boss Joshie Goldman, who eventually steals Eunice from Lenny because his economic resources and connections allow him to provide safety for her family following the Rupture, the apocalyptic event that involves the dissolution of the Bipartisan Party and the birth of a more intensely globalized United States known as "America 2.0: A GLOBAL Partnership" (322).

Shteyngart sees digital media as promoting superficiality among hybrid prospective human lovers and as presenting depravity as part and parcel of often inherently superficial human love relationships. For instance, after meeting Eunice, Lenny obsessively conducts internet searches about her, which prompt him to make superficial judgments about her and her family. He learns that Eunice's "father's business was failing" and that Eunice is the victim of child abuse, and in part he finds her victimhood attractive (38). Similarly, äppäräti equipped with Emote-Pads encourage, according to Lai-Tze Fan, superficial judgments and communication "on the level of data alone" (45). As Lenny's friend Vishnu explains just before Lenny's äppärät streams his feelings for Eunice live for mass consumption, "you look at a girl. The EmotePad picks up any change in your blood pressure. That tells her how much you want to do her" (Shteyngart, *Super Sad* 88). In turn, digital media promote corrupt behaviors in the largely superficial love relationships that emerge in the digital age, as evidenced by the fact that internet pornography readily sets the terms for individuals' perverted senses of what counts as love. For example, Eunice acknowledges what is perhaps a subconscious sense that her experience of watching internet pornography provides her with a model for her secret relationship with the older, devious, and exploitative Joshie. As she observes in a GlobalTeens message to her Korean American pen pal Jenny Kang, aka GRILLBITCH, Joshie kind of looks like "the old man who molests teens on the beach" in "Old Man Spunkers or something," one of the pornographic movies they watched as kindergarteners (226). Similarly, internet pornography helps fuel Jenny's largely superficial relationship with Gopher, her two-timing boyfriend. In an effort to both punish him for cheating on her with a "Mexican Betch" and regain his affection, Jenny visits "this new Teens site called 'D-base' where they can digitize you like covered in shit or getting fucked by four guys at once," and she sends Gopher several digitized, debasing images of herself (146, 147). As Malewitz explains, "Jenny stirs Gopher's empathy" via the images "because she is less a person than a digital object of desire—an empty vessel on which to project his equally empty fantasies" (120). Hence her immature effort to salvage her superficial relationship works, at least from her perspective, because she and Gopher have anal sex and Gopher refrains from messaging his Mexican mistress on GlobalTeens for three hours.

The kind of immaturity that Jenny exhibits endures as typical and acceptable in the world of the novel because, as Shteyngart suggests, digital culture exists for the young or for those who seek to hold onto their youth and perhaps, too, their immaturity. Indeed, Lenny, who gestures toward the digitally saturated U.S. obsession with youth and fear of old age in the novel's opening line by observing that he is "*never going to die*," confesses in his diary that his "youth has passed, but the wisdom of age hardly beckons," and his observation proves to be among his more mature acts in the novel (3, 26). Perhaps in part because his profession as a Life Lovers Outreach Coordinator (Grade G) at Staatling-Wapachung prompts him to fixate on youth by selling cutting-edge life-extension services to the rich, he shows immaturity, fixating on his physical appearance as a teenager might. He sees himself as ugly because he looks old, and he sees Eunice as beautiful because she looks like a child who is "desperately trying to coax out the preliminaries of an adult woman's body" (127). In other words, she might resemble the young Dolores Haze in *Lolita* by Vladimir Nabokov, a Russian black-humorist whom Shteyngart identifies as an influence in his interview with Elizabeth Tannen ("A Conversation with Gary Shteyngart" 176). And he behaves as a child would, too, as evidenced by the fact that he "can't clean the bathtub" (149). Along the same lines, Joshie, who reflects the U.S. obsession with youth through the cutting-edge dechronification treatments he undergoes, adopts youthful ways of being, for example by insisting that people call him by the diminutive and more youthful name *Joshie* as opposed to *Josh*; by living in a room filled with science fiction books that looks like an adolescent's; and by speaking and writing as a digital citizen such as Eunice does. And Eunice exhibits immaturity as well. As Shteyngart indicates in his interview with Tannen, Eunice is a "complex person," but the "completely digitalized world" she lives in "makes her sound the way she does": like a vapid teenager who fixates on looks and utters enigmatic and at times crude digital-age acronyms (171). She behaves immaturely, too, when she proposes marriage to Lenny early in their relationship because she feels guilty about flirting with Joshie.

Notably, because digital media cultivate illiteracy, which in turn cultivates immaturity and immature relationships among characters, Shteyngart's *Super Sad True Love Story* aims to promote maturity through literacy by overtly referencing and developing intertextual relationships with key Slavic novels. For instance, Shteyngart notes that Lenny reads Leo Tolstoy's *War and Peace*, which tells stories of both irrational and meaningful love—stories that Eunice, who "never really learned how to read texts," fails to engage with, interpret, or learn from as she purportedly seeks to develop a meaningful love relationship in her own life (277). Markedly, by virtue of her illiteracy, she lacks knowledge of the story of the young Natasha Rostova, who resembles her in her youthful beauty, and Natasha's experience of initially falling in love with the womanizing Anatole Kuragin but eventually finding meaningful love with the initially socially awkward Pierre Bezukhov, who resembles Shteyngart's Lenny and functions as Tolstoy's fictional doppelgänger (much as Lenny functions as Shteyngart's doppelgänger). Similarly, Shteyngart portrays Lenny reading Eunice passages of Milan Kundera's *The Unbearable Lightness of Being*, which depicts the exploration of different types of love by Tomas, the novel's womanizing protagonist. Tomas explores what critics characterize as heavy love, which his wife, Tereza, desires by wanting a monogamous relationship with him. And he, too, explores light love, which his mistress Sabina offers him because of her rebellious spirit. Yet Eunice by self-admission never comes to understand the meaning of Kundera's philosophical work even though erotic passages move her to participate in sexual acts with Lenny. She never comes to see the ways in which the love triangle of Kundera's novel resembles her own with Joshie (who offers her light love) and Lenny (who offers her the weight of a commitment). And, more broadly, she never gets beyond what she calls the "STATIC" of digital-age life to learn that philosophical fiction which invites contemplation might inform her overly digital life experience in useful ways (46).

Similarly, in making overt reference to Anton Chekhov's *Three Years*, Shteyngart draws attention to books' potential to guide readers in the digital age if literate readers can make meaning of them. And he suggests that it takes more than mere literacy to develop useful interpretations of a book. When Lenny as an ever-avid reader turns to his "battered volume of Chekhov's stories" in English translation to look to *Three Years* for insight into how "to overcome the beauty gap" between himself and Eunice, Shteyngart's readers see that thoughtful analysis and emotional maturity remain elusive for him (35, 36). Although Chekhov tells a story with comic elements in seemingly tragic situations that resembles the story Shteyngart narrates of Lenny's life—of the unattractive, Lenny-like Alexei Laptev's infatuation with the young, immature, and Eunice-like Yulia Sergeevna—he portrays Lenny as fixating on the wrong details of the story. According to Shteyngart, Lenny fixates on Chekhov's description of Yulia as finally convincing herself to marry Laptev because she feels her youth passing and believes there is "nothing bright to look forward to in the future" (Chekhov 354). Thus, despite his apparent dedication to fiction and print books in general, Lenny has much to learn. He fails to puzzle over complexities and nuances in the text, and he does not succeed in finding insight into his problems. He lacks the maturity to look for approaches to making love last beyond the initial romantic pursuit of a prospective partner, and never in Shteyngart's novel does he consider or mention Laptev and Yulia's long-term happiness together after their initial struggles. Lenny never mentions and perhaps never realized that each character matures in ways to make their marriage work.

As Shteyngart sees it, a mature and nonabusive relationship is ideal because it allows two hybrid Others to grow closer and make the alien more familiar through love. A model relationship consists of a thorny and transformative process that may, to borrow Lenny's idealistic words, involve partners coming to not know "where one ended and the other began" (*Super Sad* 167). But it also requires that partners respect each other's differences, which organically emerge as individuals continue to hybridize through their life experiences. Hence, until Eunice's infidelity, Lenny and Eunice appear to be building toward a model relationship. As Shteyngart portrays it, Lenny and Eunice change each other, most notably in terms of the language each comes to employ or value because of the other. For instance, Lenny, who uses dated terms such as "home-slice" and fails to "[t]alk young, live young," according to Joshie's maxim, manages to acquire new language through his relationship with Eunice, and Eunice in turn acquires that as well (223). In what is perhaps the most touching portrait of hybrid love that Shteyngart presents in the novel, Eunice calls Lenny "*kokiri*," the Korean word for elephant, and kisses his nose, and Lenny calls Eunice "*malishka*, or 'little one' in Russian," as they "feast on rice cakes swaddled in chili paste, squid drowning in garlic, frightening fish bellies bursting with salty roe, and the ever-present little pieces of cabbage and preserved turnips and seaweed and chunks of delectable dried beef" at Korean restaurants in midtown Manhattan (166). Through her relationship with Lenny, Eunice, too, attains a new reverence for language that she conspicuously fails to acquire through her relationship with Joshie, who speaks "youthfully" with her (222). She comes to appreciate introspective writing for the self of the kind that Lenny produces and values when her post-Rupture attempts at digital connection through GlobalTeens fail her. And she comes to like works of literature—even if she cannot read them herself—as evidenced by her actions when Joshie forces Lenny's eviction from his apartment in an effort to steal Eunice from Lenny. As Lenny describes it, Eunice points "to the Wall of Books out in the living room" and urges him to pack and save the books above all else (311). She shows readers that her relationship with Lenny leads her toward some semblance of emotional and intellectual growth and an appreciation of Lenny's cultural values.

Much as Lenny and Eunice change one another through their relationship, digital technology changes its partners. It emerges as a potentially enriching lover when humans morph their obsessive overreliance on it into a dynamic, purposeful, and more moderate use of it. Arguably, the thorny connection that Lenny initially has with digital-age technology exists as the most noteworthy relationship in *Super Sad True Love Story* because it survives and even flourishes in the face of apocalyptic devastation and disconnection that the Rupture brings and because it changes Lenny for the better, rendering him into a more capable and hence mature human being. By *Super Sad True Love Story*'s end, Lenny—at least according to the way he opts to tell his own story—redefines his relationship with old-media print books in general and literature in particular by way of coming to embrace digital technology and the further hybridized identity that it offers him. When he leaves New York, he changes his name to Larry Abraham (echoing the way in which Shteyngart changes his name from Igor to Gary when he moves to the United States) and eventually settles in post-Rupture Donnini in the Tuscan Free State in Italy. After unknown troublemakers purportedly breach and "pillage" his and Eunice's GlobalTeens accounts to "put together the text" that his readers see on the screen—or after Lenny orchestrates the breach after writing the diary with "the hope of eventual publication" despite his claim that he does no such thing—Larry publishes them as *The Lenny Abramov Diaries*: the GlobalTeens correspondence and digital diary that Shteyngart's readers have been reading in the form of Shteyngart's novel in print or in e-book form, depending on their preference (327). Thus, he publishes the document that showcases his evolution from Lenny into Larry by way of his relationship with digital technology—the diary that, notably, Joshie had asked employees to produce to keep track of "the mechanicals" of their "constantly changing" brains as they transformed "into entirely different people" (193).

In addition to suggesting that the act of online publication hybridizes Lenny by helping to shape him as Larry—a more savvy user of digital technology who bridges the divide between Lenny as a luddite and Eunice as a digital native—Shteyngart showcases through Larry's published diary the ways in which twenty-first-century texts can and do enrich relationships with digital technology to hybridizing ends. Notably, he posits that texts such as Lenny's diary exist as hybrid products that are born of relationships they develop with digital media because they mesh print and digital genre conventions and genres as Shteyngart's own novel meshes them. Most notably, the document exists as a love story as it might be told in an old-media print medieval romance. Yet it, too, exists as apocalyptic literature because of its portrayal of the Rupture as rapturelike—as akin to moments in the biblical book of Revelation, for instance, when what is presumably a Venezuelan navy missile sinks a ferry traveling from Staten Island to Manhattan, when activists protest and create hysteria in Tompkins Park, or when less economically privileged äppäräti users behave as the biblical left behind might behave because they fail to connect to digital networks via their devices. As Lenny describes the apocalyptic conditions, "Four young people committed suicide in our building complexes, and two of them wrote suicide notes about how they couldn't see a future without their äppäräti" (270). Moreover, the diary and Shteyngart's novel alike exist as epistolary novels of the digital age. They resemble classic epistolary works such as Fyodor Dostoevsky's "Poor Folk" or Hannah W. Foster's *The Coquette* (to cite classic Russian and American examples) and also message threads on email interfaces or print- or digital-book representations of them such as Matt Beaumont's *e: A Novel*. They contain an amalgam of GlobalTeens long-form and instant messages, showing that traditional fiction in the form of the novel can and will continue to adopt and aestheticize conventions of web writing. And they also contain digital writing of a literary quality, as evidenced by Larry's observation that "Stateside critics" agree that "Eunice Park's GlobalTeens entries" are "the gems in the text" because Eunice shows a "real interest in the world around her" (327). She writes about hybrid twenty-first-century life in a poignant voice that alternates between humorous superficial and philosophical tenors. She thereby represents with great accuracy the disconnected feeling of the postmodern, digital times that Shteyngart chronicles.

Along the same lines, Shteyngart showcases the possibilities for hybridity in literary forms through Lenny's published diary by incorporating text written in different languages. He thereby acknowledges the ways in which multilingualism has shaped his life experience and his marital relationship. He also tacitly invokes the New London Group's seminal argument that multiliteracy involves "a multiplicity of discourses" not only because of relationships humans have with new media but because of relationships they have with multilingual or linguistic Others in our "increasingly globalized societies"—relationships that, I would add, are largely possible because of digital media (New London Group 61). As a multilingual writer himself, Shteyngart sees beauty and complexity in languages of different kinds and in the experience of acquiring languages.[4] Hence he shows in fiction the hybrid beauty and complexity of multilingualism and how hybrid literature can connect apparently disparate people who, because of migration, globalization, or both, can, do, and perhaps must have relationships with one another. Early in the diary and in the novel, evidence of multilingualism in the text appears in particular through so-called broken but markedly poignant English written by Eunice's mother, Chung Won Park, who writes, for instance, that "sometimes life is suck," a phrase that Shteyngart employs as a chapter title in order to underscore the effects of globalization and migration on language (*Super Sad* 29). It likewise appears when Shteyngart includes transliterations of Russian spoken by his parents, whom Eunice and Lenny visit on Long Island. For example, Lenny mentions that his father refers to him as "[n]*ash lyubimeits*," or our favorite (132). And later in the novel, Larry includes quotations in Chinese from Cai Xiangbao, whose language is apparently dominant on the globe. According to Larry's translation, Xiangbao states that Lenny's diary exists as "a tribute to literature as it once *was*" (327; emphasis mine).

However, Shteyngart opts against showcasing multiliteracy through the multimodal interplay of image and text, suggesting that he sees visual culture as a conspicuous bedfellow of digital culture that may feed trauma rather than allow for productive introspective contemplation. In other words, although Shteyngart certainly aims for his text to serve as an example of the "resurgence of the book," which Alexander Starre describes in *Metamedia: American Book Fictions and Literary Print Culture after Digitization*, he opts against giving his book imagistic features of the kind that Starre describes (Starre 27). He opts against rendering it as a multimodal and further hybridized text such as, for instance, Mark Z. Danielewski's *House of Leaves*. And he showcases characters, including Eunice, who majors in "Images" in college, as resisting the "GLOBALTEENS SUPER HINT" to "*Switch to Images today! Less words = more fun!!!*" (Shteyngart, *Super Sad* 27). In part, this staunch commitment to words on Shteyngart's and his characters' parts spotlights the paradoxical way in which images, especially high-resolution digital ones made possible by cutting-edge technology, represent humans with great accuracy in form while dehumanizing them or underscoring their inhumane nature. Near the end of *Super Sad True Love Story*, when Lenny and Eunice attend an "art opening/Chinese welcoming party" that takes place at the Triplex, a structure that foregrounds visibility because of its transparency, Lenny feels moved by a "series of extreme satellite zoom-ins of the deadly conditions in parts of the middle and the south of our country"—high-resolution digital images of the "forced-to-be-living and the soon-to-be-dead" in postapocalyptic, post-Rupture America (317, 317, 318). He says that he thinks the photos "were amazing to see—real art with a documentary purpose" (318). Yet he also feels deeply disturbed by the images, and by one in particular: that of "the dead guy on the couch in Omaha" and "a German shepherd shot point-blank," her "puppy of negligible weeks, maybe days" near her "still-swollen teats" (320, 321). Although Lenny never fully understands why this image worries him so profoundly, Shteyngart's readers see that the image shows, with unsettling accuracy, the dark underbelly of human nature. Much as the artist's "gimmicky" shooting spikes make Lenny feel as though he is "not a human being" when he attempts to approach the artist, this artist's images accentuate the inhumanity of humans (319).

Ultimately, Shteyngart portrays Larry as a champion of language and literature who counters visually designed acts of dehumanization with verbal activist articulations that make a more lasting and meaningful difference than activist acts that precede the Rupture. In the final scene of the novel, while Larry visits his friends Giovanna and Paolo at their Italian country home, two actresses who have "just been charged with playing Eunice Park in a new Cinecittà video spray" of Larry's diaries arrive (330). As practice for their soon-to-be digitized performance, they begin to mock Eunice and Lenny as they know them—as characters in a text. They fail to understand that reality and artistic renderings interconnect in manifold and complex ways and that humor is subjective and has its limits. One of the actresses pulls "at her eyelids" and sticks out her "upper front teeth" in racist mockery (330). And then she launches into what Larry calls a "fairly accurate rendition of a spoiled pre-Rupture California girl" (330). Meanwhile, the other actress falls "prostrate on the ground at her feet, weeping hysterically" in an insensitive portrayal of Lenny and his sincere love (330). Impressively, in the face of the cruelty that the actresses exhibit via the visual performance art they produce, Larry presents a "subdued mien," sets his "mouth into its own version of Eunice's dead smile," musters a mechanical laugh, and then undercuts the dehumanizing artistic acts of the actresses by virtue of writerly cunning and creativity, which he has always revered even as they have eluded his shifting identity—even though who Larry was and who Larry is remain ambiguous (330). Lenny tells, in spoken form, a story that merges truth and lies into a hybrid whole and that is therefore evocative of the rhetorical form of his diary, a document that counterbalances sincere or true representations of love's thorny nature with lies as the inevitable fabric of fiction. In it, he to some degree achieves the mature realization to which he comes by novel's end: the realization that he is "*going to die*" and not live forever as he intimates in the novel's opening (304). He tells the actresses that Lenny and Eunice "didn't survive" even though in many ways they do, for instance because Lenny becomes Larry and because the characters live on in art (331). Notably, he tells the actresses a story that is "more gruesome than any of the grisly infernos splashed" in image form "on the walls of the neighboring cathedral" and evocative of the apocalyptic tone of the unnamed artist's perplexing digital photographs (331).

However, Shteyngart's readers never encounter Larry's harsh details as viewers of the unnamed artist's photographs encounter analogous details in imagistic form. Thus, they have a hybrid and tempered relationship with Larry's fiction, which manifests within Shteyngart's fiction. For readers, Shteyngart splits the difference between the "silence, black and complete" that Larry says he desires and the experience of hearing the purportedly shocking tale (331). And by novel's end, readers have control of how they imagine Lenny's and Eunice's deaths. They function much as the "tableau of olive trees and grain fields, arrested by winter" do in the Italian countryside in the novel's closing moments as they dream of "a new life" (331). More significantly, they are left to contemplate the value of fiction in its various and increasingly hybridized and flexible forms. And they are left to contemplate the value of Shteyngart's novel as a twenty-first-century work of fiction that hides behind the guise of a super sad and purportedly true love story that in reality merges truth, whatever truth may be, with uplifting and satirical moments that by design counter Shteyngart's title as a label. Indeed, readers see that through meaningful relationships with fiction, authors and readers alike can stimulate their imaginations. They can portray or read about death, which Lenny artfully depicts to protect himself from cruelty. But they can also imagine through writing and reading a more humane human race that appears as apparently elusive in the fictionalized future of Shteyngart's text and in the notably similar contemporary reality out of which it emerges. And they can perhaps imagine otherwise unknowable aspects of hybrid Others to themselves to foster meaningful connections and perchance true love among increasingly hybrid real-life individuals. They can even begin the process of rethinking and remaking the globalized and apparently ever-digitizing twenty-first century—a century that seeks to undercut the printed word and thought-provoking words in general because of their capacity for adapting and sustaining an undeniable and enduring influence.

Digital Screens, Human Relationships, and the Problem of Misreading in Kristen Roupenian's "Cat Person"

In "Cat Person" (2017), the focus of the second part of this chapter, Kristen Roupenian portrays the emergence of twenty-first-century romance as the result of the interplay of in-person and digital communication much as Shteyngart does. In Roupenian's story, both of these types of communication take on unsettling features because prospective twenty-first-century lovers seem more acclimated to interacting with digital devices than with one another. For Roupenian, in-person communication is awkward in an array of ways because individuals who spend their days interacting with screens lack a full sense of the inherent complexity of prospective lovers. For instance, Robert and Margot, the story's protagonists, both engage in acts that are markedly reductive in nature when they meet in a notably screen-mediated and fantasy-oriented space: at the "artsy" downtown movie theater where Margot works at the concession stand and where Robert makes solitary trips to watch a movie screen that functions as an antecedent to the digital screens that pervade twenty-first-century life (77). In their interaction with each other, they reduce the other to some fragment of a whole character instead of attempting to develop a literacy of the other that would function as a foundation for a meaningful romantic connection. For instance, the narrator reveals that Margot reduces Robert to a mode of entertainment akin to a movie screen. She flirts with Robert as she does with many customers—as a "habit" that helps her avoid boredom at work in late-capitalist America (77). Similarly, Robert reduces Margot to her unfulfilling position, referring to her as "[c]oncession-stand girl" when he asks her for her phone number (78). He talks to her much as the narrator of Herman Melville's "Bartleby, the Scrivener" talks to his Wall Street employees, reducing them to idiosyncrasies by virtue of nicknames such as Ginger Nut, a worker who brings ginger nut cakes to the office (78).

Roupenian suggests that the knowledge of each other that prospective lovers acquire in the digital age is faulty because digital devices allow prospective lovers to reduce themselves to idealized characters through rhetorical inclusions and exclusions. She identifies digital communication as being dominated by constant wit, which David Foster Wallace, for instance, critiques in his portrayal of postmodernism in "My Appearance," a fictional work that, like the nonfictional "E Unibus Pluram: Television and U.S. Fiction," articulates Wallace's New Sincere paradigm. Much as Wallace critiques postmodern irony through a representation of an absurd web of irony, which Edilyn as Wallace's sincere heroine encounters in "My Appearance," Roupenian critiques the "elaborate scaffolding of jokes via text" that unfold following Margot and Robert's initial in-person meeting (78). As the narrator describes it, Robert was "very clever," and Margot "found that she had to work to impress him" (78). Yet the substance of their conversation is the antithesis of substance, as evidenced, for instance, by Margot's reliance on the internet as a means by which to move the conversation forward. According to the narrator, Margot would "think of something funny to tell him or she'd see a picture on the internet that was relevant to their conversation, and they'd start up again" (78). As a result, Margot "still didn't know much about him, because they never talked about anything personal" despite the existence of a peculiar "exhilaration" to their digital communication as a result of telling each other "two or three good jokes in a row" (78). Hence their relationship exists in a paradoxical state in that they consistently spend time alone together, to reference the title of Turkle's third book on computers and people and the social problem that the book explores. When the two meet again in person—at a 7-Eleven—Robert greets Margot "without ceremony, as though he saw her every day," because in a sense he does see her—or at least a part of her—daily through the messages they exchange (79). They have the illusion of intimacy even though neither knows much about the other's identity. And they resemble Lenny and Eunice, who fail to attain knowledge about each other early in Shteyngart's novel. Digital screens screen each from the other while purportedly connecting them.

Roupenian intimates that the paradoxical condition that digital-age lovers inhabit—a condition in which they both know one another and utterly lack knowledge about one another—fuels the formation of assumptions, stereotypes, and categorizations of behavior that are largely rooted in screen-mediated representations of reality, which complement the false reality that text messages create. Notably, Roupenian's title, "Cat Person," alludes not only to Robert's claim (on two occasions) that he has cats and the fact that Margot has cats, but to catfishing (i.e., the act of using a fake online persona to lure a lover into a relationship) and to a Hollywood-born and wholly dualistic way of thinking about men and women in love. As movies such as *The Truth about Cats and Dogs* (starring Uma Thurman and Janeane Garofalo) or *Must Love Dogs* (starring John Cusack and Diane Lane) suggest, Hollywood screens tell the masses that cats and dogs function as metaphors for lovers or prospective lovers. On one hand, men function as dogs and women function as cats. On the other hand, individuals—be they men or women—function as dog people or as cat people, which Robert in Roupenian's story purports to be and which Margot is. In the way that romantic comedies bolster these ways of thinking in categories, online dating sites, too, bolster them, as demonstrated, for instance, by profile questionnaires that these sites invite users to complete. Although such sites claim that their algorithms for matching prospective lovers based on answers they provide to questions about preferences, habits, and identity features are effective, in reality they fail to reveal or analyze the complexity of human character or deduce compatibility accurately. As Eli J. Finkel and Susan Sprecher explain in their 2012 *Scientific American* article, "The Scientific Flaws of Online Dating Sites," "sites such as Chemistry.com, PerfectMatch.com, GenePartner.com, and FindYourFaceMate.com have claimed that they have developed a sophisticated matching algorithm that can find singles a uniquely compatible mate," but these "claims are not supported by any credible evidence." In other words, these sites allege that the flattened identities which prospective lovers encounter via their screens will lead to love, but love remains elusive as long as oversimplifications of human character masquerade as viable building blocks in the process of finding love.

The assumptions, stereotypes, and categories toward which Roupenian's title gestures pepper her text, suggesting that she sees lovers as making viral in their real-life experiences unsettling reductions of identity that serve as the foundation for problematic relationships in the digital age. For instance, the narrator describes Margot as consistently struggling against assumptions that Robert makes or threatens to make about her—at least according to her speculations about his thinking. And Margot likewise struggles against stereotypes Robert ascribes or threatens to ascribe to her. According to the narrator, Robert allows a "secret drama" to develop in his imagination while Margot is home in Saline for spring break (93). This drama involves her reconnecting with a high school boyfriend even though, in a reality that Robert never comes to know, Margot's high school boyfriend is gay and not interested in a romantic reconnection. When Margot returns to college from spring break and challenges Robert's idea that they see a film about the Holocaust on their first date, which is notably screen mediated, Robert again makes an assumption about Margot: he jokes about "how he was sorry that he'd misjudged her taste" and, in a derogatory way, offers instead to buy tickets to a romantic comedy of the kind that influences Roupenian's title (82). Similarly, Margot feels anxious about ordering a drink at the bar that admits her as an underage drinker without an ID in part because she worries about perceptions of her identity. She knows relatively little about alcoholic beverages and fears that the cheap beers she orders lack cultural capital in Robert's eyes. As the narrator declares, "at the places she went to, they only carded people at the bar, so the kids who were twenty-one or had good fake I.D.s usually brought pitchers of P.B.R. or Bud Light back to share with the others" (84). And as the narrator elaborates, "She wasn't sure if those brands were ones that Robert would make fun of" (84). Finally, after Margot tacitly propositions Robert to leave the bar for sex and they contemplate their destination, Margot feels judgment from Robert about her identity as a college student who lives in the dorms, which they avoid as a destination because of Margot's roommate. To reference the narrator's words, Margot imagines that Robert perceives her student identity as "something she should apologize for" (86).

Roupenian positions Robert as similarly vulnerable to Margot's stereotypes and assumptions, suggesting that Roupenian sees stereotyped women in the twenty-first century as guilty of perpetuating stereotyping and assumption making as acts—even if they have self-preservation as an objective in doing so. Most significantly, in assuming that Robert wants to murder her en route to the movie theater on their date, Margot consistently engages in a fantasy rooted in the apparent experience of watching TV news footage about real-life violence against women and horror movies that are inspired by newsworthy events. She fears violence of the sort that looms large in Roupenian's "Look at Your Game, Girl." As the narrator describes it, "Before five minutes had gone by," Margot "became wildly uncomfortable, and, as they got on the highway, it occurred to her that he could take her some place and rape and murder her" (80). Interestingly, Robert anticipates the fantasy that plays out in Margot's mind because he shares her cultural knowledge of horrific narratives of violence against women. Robert randomly affirms that he has no plan to rape or murder Margot, and after he makes this remark, the narrator exposes Margot's thoughts about herself as a stereotypical woman who makes stereotypical assumptions about men. According to the narrator, Margot "wondered if the discomfort in the car was her fault, because she was acting jumpy and nervous, like the kind of girl who thought she was going to get murdered every time she went on a date" (81). Similarly, in an absurdist moment in Roupenian's narrative, Margot makes assumptions about Robert's identity and her compatibility with him based on broad categories that online dating sites highlight in users' profiles. When she arrives at his apartment, the narrator observes that she tempers her fear of experiencing violence at his hands by focusing on "evidence of his having interests that she shared, if only in their broadest categories—art, games, books, music" (87). As the narrator notes, these broad categories strike Margot as a "reassuring endorsement of her choice" to sleep with him—even though any and every modern-day American may well like art, games, books, and movies (87). She sees his interests as informative even though readers see them as laughable because they reveal virtually nothing about Robert other than the fact of his modern-day humanity.

In ways echoing Shteyngart, who views screen culture as endorsing superficiality and depravity, Roupenian suggests that elements of screen culture that influence her characters' perceptions of one another come to render digital-age reality as surreal and also as akin to a movie in problematic ways. Repeatedly, Margot perceives her unsettling sexual encounter with Robert in cinematic terms, and Robert may perceive it as such as well, though readers never know for certain because the narrator's close-third-person narration reveals only Margot's perspective. Although she feels "confused" by his use of digital media at the start of their sexual encounter—when he opens his laptop to play music—Margot involves media in their encounter in deeper and more unsettling ways by virtue of conceiving of Robert as imagining a bad pornographic film that he might watch online (88). As the narrator puts it, "During sex," Robert moved Margot "through a series of positions with brusque efficiency, flipping her over, pushing her around," making her feel like "a doll made of rubber, flexible and resilient, a prop for the movie that was playing in his head" (91). Along the same lines, Margot sees herself in the sexual encounter as though she is simultaneously acting in and watching a movie that contains a high-angle shot. As the narrator explains, even though Margot feels that "her last chance of enjoying this encounter" disappears and even though she feels "a wave of revulsion" at the sight of the overweight Robert that "might actually break through her sense of pinned stasis," she opts to "carry through" with sex when she imagines "herself from above, naked and spread-eagled with this fat old man's finger inside her" (91). The narrator continues, noting that "her revulsion" turns to "self-disgust and a humiliation that was a kind of perverse cousin to arousal" as a result of her cinematic vision (91). Ultimately, Margot proceeds with consensual yet undesired sex, an act that at best functions as a mistake or at worst showcases the hazy realities of consent and the peculiar power dynamics at play between an older, presumably white man and a younger, stereotypically pretty, college-educated, presumably white woman.

As a result of engaging with reductions of identity and screen-mediated perspectives, lovers inevitably misread one another's actions, a noteworthy occurrence particularly because of existing criticism of digitization's transformative if not negative effects on reading.[5] For instance, Robert misreads Margot's flinch during sex, thinking that it signifies her virginity as opposed to her displeasure with his actions. He fails to inquire about Margot's sexual history: that the loss of her virginity "had been a long, drawn-out affair preceded by several months' worth of intense discussion with her boyfriend of two years, plus a visit to the gynecologist and a horrifically embarrassing but ultimately incredibly meaningful conversation with her mom" (90). In turn, Margot misreads Robert—or at least she fears that she misreads him. According to the narrator, when Margot ruminates over Robert's choice of a film about the Holocaust for their first date, "a totally different interpretation of the night's events" occurs to her (82). As the narrator elaborates, "She wondered if perhaps he'd been trying to impress her by suggesting the Holocaust movie, because he didn't understand that a Holocaust movie was the wrong kind of 'serious' movie with which to impress the type of person who worked at an artsy movie theatre, the type of person he probably assumed she was" (82). Moreover, Margot may well misread Robert's sexual history. Because Robert is older than she is, she assumes that he must have more sexual experiences. She assumes that he "must have seen more breasts, more bodies" than her past lovers had seen (89). And she altogether fails to entertain the possibility that the older and unattractive Robert may lack sexual prowess because he lacks experience with women, as evidenced, for instance, by the narrator's mention of the fact that he is unable to unhook Margot's bra.

Misreadings of reality give way to the utter dissipation of reality in Roupenian's story as characters in the text feel the detrimental effects of lies and simulations, both of which Shteyngart explores at the end of *Super Sad True Love Story* and both of which Roupenian sees as setting the terms for everyday lived experiences in the digital age. Perhaps most obviously yet least consequentially, Robert's claim that he owns two cats may be a fiction because neither Margot nor Roupenian's readers ever see any cats. The narrator draws readers' attention to this detail, noting that Margot remembered that Robert had "talked a lot about his cats and yet she hadn't seen any cats in the house, and she wondered if he'd made them up" (95). More consequentially, the setting of Roupenian's story highlights the more pervasive and less visible nature of simulation as the norm in twenty-first century life, particularly for young and privileged Americans who use digital devices and media most. Indeed, Margot as a young and economically privileged American and Robert as an older and apparently less privileged one come from different worlds figuratively. But they likewise come from different parts of a town that by nature amplify the contemporary interplay of reality and simulation, if not the unsettling deconstruction of reality and simulation as a binary. Margot inhabits and sustains a literacy of the purportedly fake if not fantastical "student ghetto" of the college town, which illustrates in fictional form the veracity of a dictum that circulates among Ann Arbor's residents (82). According to it, Ann Arbor is twenty-seven square miles surrounded by reality. She functions as part of what Blake Gumprecht in *The American College Town* identifies as a definitive feature of college towns: a *"transient"* population (10). She never inhabits the town in a permanent, meaningful, or real sense. She arrives on campus in the fall and leaves during spring, winter, and summer breaks. And the narrator's description of the "big multiplex just outside town" as opposed to the downtown artsy theater where Margot works spotlights a paradox of the college town—a peculiar inversion or dissolution of the real and the fake (80). Paradoxically, the corporate movie theater, a mass-produced, late-capitalist replacement of downtown theaters of old, is more real than the original on which it is based because students who give the town a fake feel "didn't go there very often" (80). The fake, new theater is more real than the old theater downtown because it exists beyond the bounds of the Neverland-like world where transient college students live. It exists mostly for the town's permanent or real residents, who seek to avoid the crowds of college students in the downtown area.

The dissipating distinction between the fake and the real that the setting of Roupenian's story showcases reflects the dissolving line between lies and fantasies, on one hand, and truth and reality, on the other, in the digital-age human psyche. Although individuals have turned to fantasy as a means by which to fuel romance and learned about lovers or prospective loves through mediated means such as love letters throughout modern human history, Roupenian represents fantasy, a genre that she explores in works such as "The Good Guy" and "Scarred," in nearly grotesque ways. And she does so to underscore its potential toxicity, especially when lovers come to prefer fantasy to reality. For instance, Roupenian represents Margot as turning to a new fantasy when her initial one goes awry. Following her wholly unsatisfying sexual experience with Robert and upon realizing that the real-life Robert is nothing like the man she came to imagine based on their text-message exchanges, Margot "imagined that somewhere, out there in the universe, there was a boy who would think that this moment was just as awful yet hilarious as she did, and that sometime, far in the future, she would tell the boy this story" (92). As the narrator continues, however, exposing the problem with Margot's thinking, "there was no such future, because no such boy existed, and he never would"—presumably because cisgender heterosexual men who appeal to Margot lack the capacity to fully understand, let alone empathize with, cisgender heterosexual women's experiences of dating (92). Along the same lines, after ignoring Robert's text messages in the days following their unsatisfying sexual encounter and in an effort to suppress the reality of that encounter, Margot comes to long for fantasy because her reality is dissatisfying. In the narrator's words, Margot "would find herself in a gray, daydreamy mood, missing something, and she'd realize that it was Robert she missed, not the real Robert but the Robert she'd imagined on the other end of all those text messages during break" (95–96). She longs for a Robert who never has existed and never will exist in the real world.

Because digital citizens as Roupenian represents them prefer fantasy to reality, they exist as incapable of understanding or processing realities that confront them—even though, paradoxically, they believe they crave what is real and true. For instance, the narrator suggests that Margot believes herself to crave truth in observing that Margot likes the "stunned and stupid" look on Robert's face just before they have sex (89). In the narrator's words, "maybe this was what she loved most about sex—a guy revealed like that" (89). However, when she meditates on the reality of her unfulfilling sexual encounter with Robert, she expresses a desire to utterly remove him and her relationship with him from her life and reality. The narrator describes Margot's postcoital actions once she arrives in the comfort of her dorm room, noting that she "slept for twelve hours, and when she woke up she ate waffles in the dining hall and binge-watched detective shows on Netflix and tried to envision the hopeful possibility that" Robert "would disappear without her having to do anything, that somehow she could just wish him away" (95). She turns to a screen in an effort to screen out an undesirable and perplexing yet undeniable part of her very real sexual history. Moreover, Margot feels overwhelmed by the way in which text messages from Robert remind her of the reality she seeks to deny. She initially opts against replying to his messages in large part because she has no sense of how to explain that she wants to break off their relationship. And eventually, after Robert replies to a breakup message that Margot's friend Tamara writes and sends on Margot's behalf without Margot's permission, Margot asks Tamara to read Robert's message, presumably because she feels incapable of confronting the reality it represents. Margot even attempts to erase history by generating fiction in its place. As the narrator explains, Margot's friend Albert and many of her other friends had "heard a version of the story, though not quite the true one" (97). They hear a story that positions Robert as the villain and Margot as the victim of what in reality exists as a much hazier narrative of a screen-mediated relationship gone awry.

Perhaps inevitably, the truth of the self functions as the most noteworthy truth that digital citizens fail to process in the frame of Roupenian's story. In other words, digital citizens as Roupenian represents them lack a full understanding of themselves and their motivations, as demonstrated, for instance, by Margot's limited critical awareness of her own harsh judgments, superficial values, and narcissistic tendencies, the last of which resemble those expressed by characters at various points in Shteyngart's novel. Although Margot briefly entertains the notion that she is "being unfair to Robert, who really had done nothing wrong, except like her, and be bad in bed, and maybe lie about having cats," she predominantly opts to indulge or overindulge in appreciation of herself and her own superficial attributes over the course of Roupenian's story—even though she paradoxically fails to value herself enough to avoid having sex with a man who disgusts her (97). The narrator notes that Margot sees herself as "something precious" when Robert kisses her forehead for the first time (79). And she sees herself as a "delicate, precious thing" when he kisses the top of her head later in the story (83). She likewise consistently sees herself as beautiful, as evidenced, for instance, by the way she sees herself in the reflection of Robert's eyes and imagines herself through his eyes as they stand outside in the wintry college town. As the narrator puts it, "she could see how pretty she looked, smiling through her tears in the chalky glow of the streetlight, with a few flakes of snow coming down" (83). She may not be aware of herself as desiring power, but she seems to desire it because her gender by default denies it to her in patriarchal American society, and she employs her looks and her femininity to obtain some semblance of power and control—or at least the illusion of it. She may not identify as being manipulative, but she exploits her own identity as a weak young woman by speaking "self-deprecatingly" about herself and revealing her own purported anxieties with the strategic goal of rendering Robert into "a large, skittish animal, like a horse or a bear, skillfully coaxing it to eat from her hand" (85). And whether her perception of Robert as animal-like is accurate or not, she likes thinking of him as "hungry and eager to impress her" (85). She likes "the obedient way" he follows her out of the bar (86). And she feeds her own overly healthy ego by allowing herself to be "carried away by a fantasy" of "pure ego" (89). In her fantasy, she imagines Robert as enthralled by her beauty and fails to understand herself as egotistical. She imagines that he thinks that she is "so perfect, her body is perfect, everything about her is perfect, she's only twenty years old, her skin is flawless, I want her so badly, I want her more than I've ever wanted anyone else, I want her so bad I might die" (89).

In addition to representing digital citizens as lacking a critical literacy of themselves, Roupenian represents digital citizens as natural-born cyborgs, which Andy Clark theorizes in *Natural-Born Cyborgs: Minds, Technologies, and the Future of Human Intelligence*. As natural-born cyborgs, Roupenian's characters merge with technology without implantation much as the digital natives of Shteyngart's novel do. And their profound mergers leave them functioning more like machines, as evidenced by the fact that they fail to understand and process a full range of human emotions in mature ways. Consistently, Margot struggles to manage her own anxiety, for example when Robert fails to respond to her texts in a timely manner early in their relationship. In the narrator's words, "when she asked him a question and he didn't respond right away she felt a jab of anxious yearning" (79). And she also feels anxiety because of her youthful inexperience, as illustrated by the narrator's remark that she "actually was a little anxious about what to order" at the bar (84). Similarly, Margot struggles to manage her own anger after having sex with Robert—even though or perhaps because the source of it exists as enigmatic to her. She feels angry with herself for sleeping with a man whom she did not desire, though neither she nor the narrator ever pinpoint the cause of the "black, hateful aura" that Margot emanates (93). She fails to process disgust with herself as well, especially when Robert kisses her forehead, making her feel not at all precious this time but "like a slug he'd poured salt on, disintegrating under that kiss" (94). Finally, she fails to process fear and dread. When Robert sends her messages after they have sex, she worries that the messages will "keep coming and coming; maybe they would never end" (95). And she feels "sick and scared" when she sees him at the bar she frequents after their relationship ends, though likely not because she feels that he poses a physical threat to her (97). She feels this way instead because she fails to process the reality of the development and dissolution of their relationship. She is unable to make sense of the clear emotional disconnection between herself and Robert, who underscores the problem of disconnection in a text to Margot. As Robert puts it, "*I felt like we had a real connection did you not feel that way or...*" (98).

Despite their profound mergers with digital devices, twenty-first-century citizens as Roupenian represents them exist as notably different from cyborgs as Haraway theorizes them in "A Cyborg Manifesto." Whereas Haraway argues in her manifesto that technology is political and that the cyborg functions as a metaphor for how Americans can realize socialist and feminist ideals through celebrating cyborglike mergers and mixtures, Roupenian postulates that the natural-born cyborgs in her text lack an awareness of the political implications of their actions and inactions—even though Roupenian's story as a rhetorical work of art pointedly underscores the social problems that apolitical thinking about identity in general and the self in particular perpetuates. Roupenian, too, posits that the natural-born cyborgs of her tale make manifest the antithesis of solidarity-based social change. Characters in her story showcase utter disconnects between their thoughts and feelings, on one hand, and their actions on the other, thus rendering their actions as markedly different from the purposeful liberatory work of social movement organizers and activists. Although Roupenian never allows her readers to obtain a clear sense of Robert's thoughts because of the way in which she aligns the narrator's perspective with Margot's, she hints at ways in which Robert's feelings and actions are disconnected from one another and perhaps in part function to produce his chauvinism. For instance, he tells Margot that she does not need to apologize for laughing at his question about her virginity, but as the narrator suggests, "she could tell by his face, as well as by the fact that he was going soft beneath her, that she did" (91). Likewise, he exhibits a disconnect between his thoughts and actions during and after sex with Margot. According to the narrator, Robert appears to have some semblance of sexual dysfunction because "when he was on top" of Margot "in missionary, he kept losing his erection, and every time he did he would say, aggressively, 'You make my dick so hard,' as though lying about it could make it true" (91–92). Along the same lines, there exists a disconnect between Robert's treatment of Margot during sex and his postcoital affection toward her—when he strokes her hair and trails "light kisses down her shoulder, as if he'd forgotten that ten minutes ago he'd thrown her around as if they were in a porno" (92–93).

More consequentially, Roupenian invites readers to contemplate the disconnect between Margot's feelings and actions. She invites them to consider why Margot has sex with Robert even though she feels repelled by him. Hence she prompts readers to consider the social pressures on women to behave politely to the point of experiencing violation instead of expressing their own desires or lack thereof and, in turn, advocating for their own and other women's rights in a patriarchal American nation. Throughout the story, Margot piles disconnect upon disconnect in a digital age of apparent hyperconnection, for instance when she feels disgusted by Robert's "terrible" and "shockingly bad" kiss yet develops a "tender feeling toward him again, the sense that even though he was older than her, she knew something he didn't" (84). She feels no desire to engage in sex with Robert but proceeds with it anyway because "the thought of what it would take to stop what she had set in motion was overwhelming" and because she does not want to "seem spoiled and capricious, as if she'd ordered something at a restaurant and then, once the food arrived, had changed her mind and sent it back" (88). And she wants to laugh at his crass pillow talk but instead opts to "smother her face in the pillow to keep from laughing again" (91). After sex, she sees "the perfect opportunity to send her half-completed breakup text," which reflects her feelings of wanting to end things with Robert, "but instead she wrote back, *Haha sorry yeah* and *I'll text you soon*, and then she thought, Why did I do that? And she truly didn't know" (96). Finally, she tells her friend Tamara that Robert is "a nice guy, sort of," even though she secretly wonders "how true" her statement is—how well what she says represents the truth of what she feels (96).

Roupenian leaves little hope for humans to more meaningfully read their own feelings and translate them into actions. And as Roupenian seems to see it, the inevitable consequence of this persistent disconnection between thoughts or feelings on one hand and actions on the other is quasidystopian social collapse, though of a less literal sort than Shteyngart portrays. For Roupenian, this social collapse emerges in part through the paradigm of a digital-age linguistic collapse. To appropriate the tacitly critical words of the narrator of "The Boy in the Pool," instead of making the world a futuristic and "fantastical place" where social and digital media magically bring human fantasies to life, digital devices as Roupenian represents them by the end of "Cat Person" bring about human regression, a notion that Roupenian illustrates through her description of Margot in the story's closing scene (163). According to the narrator, Margot "[c]urled up on her bed with Tamara" with "the glow of the phone like a campfire illuminating their faces" (97). She resembles an early human in a fire-lit cave. And in reading messages that arrive from Robert, she internalizes language that is evocative of modern-day notions of primal human communication. Robert's text messages lack punctuation and contain misspellings. For example, he writes, "*Hey maybe I don't have the right to ask but I just wish youd tell me what it is I did wrog*" (98). He turns to digital-age shorthand that involves replacing words with single letters, asking her, "*When u laguehd when I asked if you were a virgin was it because youd fucked so many guys*"

(98). And he engages in harassment, sending the same question without a question mark repeatedly. He asks, "*Are you fucking that guy right now*" and then asks "*Are you*" three times in apparently quick succession when Margot opts against replying (98). Finally, he showcases the seeming pinnacle of his own regression and what Roupenian, in her interview with Deborah Treisman, says she sees as "unequivocal evidence about the kind of person he is" by calling Margot a "*Whore*" in the final text message and the final word of the story ("Kristen Roupenian on the Self-Deceptions"; "Cat Person" 98). He resorts to a vulgar insult that is evocative of simplistic, dualistic ways of thinking about women: as mothers or as whores who exist only to satisfy male needs and desires, as variations on the ones and zeros of binary code. And he resorts to communicating his insult through a machine, leaving readers to wonder why twenty-first-century humans subject themselves to abuse at the hands of inanimate objects that they could exert control over even though they instead appear to be controlled by them.

Ultimately, Roupenian's story functions as a literary parable for the digital age. Tacitly and by virtue of its content, form, and reception, it showcases its own value as literary fiction. And in accord with the rhetorical purpose of Shteyngart's novel, it showcases the value of literary fiction in general, a point toward which Constance Grady gestures in her 2017 *Vox* article, "The Uproar Over the *New Yorker* Short Story 'Cat Person,' Explained." As Grady puts it, "Regardless of whether or not 'Cat Person' is a great short story or just an okay short story, whether it's deeply subversive or highly problematic, it has been exciting to see the cultural discourse revolve around a short story for a spell," because attention to Roupenian's story is "a reminder of how immensely powerful and valuable fiction can be, and why it's worthwhile to pay attention to it and learn from it." Of course, Roupenian's hybridized story, which both represents and (in its publication forms) bridges digital and physical worlds, is particularly valuable because it attempts to teach readers more than a mere lesson about the risks involved with dating in an American society that allows rape culture to run rampant. It also attempts to teach more than a simple lesson about the dangers of digital communication. And finally, it teaches more than a lesson about the complexities of privilege or the value of feminist ideals and solidarity. Instead, it invites thoughtful reflection on the surreal and volatile intersection of digitization, gender, privilege, and sex and on the intersections of our own identities as twenty-first-century citizens. Indeed, through carefully crafted prose, Roupenian's story invites readers to cultivate a literacy of everyday approaches to avoiding hyperrealistic dystopian features that complement Shteyngart's fantastical dystopian events. Thus, Roupenian's story invites readers to realize that both realistic and fantastical literary fictional works remain very much worth their cultural capital if they seek to create positive social change in the world.

CHAPTER 2

Searching History in Thomas Pynchon's *Bleeding Edge* and Jennifer Egan's *A Visit from the Goon Squad*

The onset of the digital age brought with it not only transformations in human relationships, which Gary Shteyngart and Kristen Roupenian explore, but also a radical transformation in conceptions of time and human relationships with time, which Thomas L. Friedman intimates via the title, circumstances of publication, and controlling metaphor of *The World Is Flat*. In publishing his lengthy book as a so-called brief history in 2005, just four years into the twenty-first century, and also in suggesting that the world morphed quite quickly while he "was sleeping," Friedman showcases the peculiarities of time that define the times and that change the ways in which humans conceive of history as a result of digitization (8). He showcases the notion that Americans feel not only altered notions of space, which Friedman points toward when he observes that "Globalization 3.0 is shrinking the world from a size small to a size tiny," but a sense of lost time of the sort that Judy Wajcman theorizes in *Pressed for Time* (Friedman, *The World Is Flat* 10). According to Wajcman, whose work evokes the anxiety about time that appears in Henry Adams's *The Education of Henry Adams*, "There is a widespread perception that life these days is faster than it used to be" even though "modern machines [were] supposed to save, and thereby free up, more time" for their users (1). Instead, as Wajcman indicates, "the iconic image that abounds is that of the frenetic, technologically tethered, iPhone- or iPad-addicted citizen" (2). And this citizen expounds upon notions of hybridity that theorists such as Donna Haraway in "A Cyborg Manifesto" or Homi K. Bhabha in *The Location of Culture* put forth. Thus, according to Wajcman, the masses see "[r]apidly evolving information and communication technologies" as "marking a whole new epoch in the human condition" (2).

This chapter addresses these changing notions of time through analyses of Thomas Pynchon's *Bleeding Edge* and Jennifer Egan's *A Visit from the Goon Squad*, twenty-first-century literary works that (to make a metaphor of the search histories that web browsers create) search history and conceptions of history in order to develop meaningful, countercorporate, and counterexceptionalist visions of the digital-age American future. Perhaps because he studied engineering physics and had the formative experience of working at the Boeing Corporation after completing a stint in the navy and an English degree from Cornell University, Thomas Pynchon, the subject of the first section of this chapter, shows a unique concern for historical developments in technology and the politics of those developments. As early as 1966, when he published *The Crying of Lot 49*, he makes reference to "[o]nes and zeroes," digital-age binary code, which is in ways representative of the depthless and dualistic American thinking that he criticizes in the novella (150). In turn, he valorizes counterculture and paranoia, the latter of which he sees as a means by which to make connections. As the narrator of *Gravity's Rainbow* suggests, "there is something comforting—religious, if you want—about paranoia" because it offers connectivity that juxtaposes with the antiparanoid individual's notion that "nothing is connected to anything, a condition not many of us can bear for long" (441). Indeed, in *Gravity's Rainbow*, which primarily takes place between 1944 and 1945, Pynchon focuses on the moment of modern technology's birth and V-2 rockets as an early example of modern technology's violent potential. In the novel, he asks readers to contemplate whether the rocket will "enable humanity to go to the stars—or to destroy itself," to cite David Cowart's words (*Thomas Pynchon and the Dark Passages* 12). And this question can be adapted to address the flat world that Pynchon scrutinizes in *Bleeding Edge*. As a meditation on the historical significance of the 9/11 terrorist attack and arguably Pynchon's most robust consideration of modern technology, *Bleeding Edge* hypothesizes that the internet resembles the rocket. It is a brainchild of the Cold War that might provide Americans with an opportunity for transcendence or self-destruction. Furthermore, the internet and the flat world that it creates permeate deep into American consciousness to shape and reshape conceptions Americans have of history and the future at the start of the twenty-first century, a symbolic moment in which Americans have an opportunity to change the course of history.

My analysis of *Bleeding Edge* considers the novel as a literary philosophy of digital technology that extends ideas Pynchon presents in *The Crying of Lot 49* and in his 1984 *New York Review of Books* essay, "Is It O.K. to Be a Luddite?" Emerging out of Amy J. Elias's argument that Pynchon's novels "imply a philosophy of history" and also from Inger H. Dalsgaard's argument that "[s]cience and technology simultaneously provide countless occasions for Pynchon to comment on historical, political and social issues," I argue that the novel positions protagonist Maxine Tarnow as a twenty-first-century version of Oedipa Maas who, in her investigation of dot-com CEO Gabriel Ice's finances, scrutinizes the position of the media and digital technology as driving forces in the production and nonproduction of American history before and after 9/11 (Elias 124; Dalsgaard 165). As Pynchon intimates, media set the terms for human thinking about history and enable a widespread and problematic American nostalgia—most notably for the 1990s and for Cold War–era ways of thinking and being that existed prior to the fast-paced era that Wajcman analyzes. This nostalgia leads Americans to enter into unsettling cycles of history that counter Pynchon's previous conceptions of history as entropic and that allow for the erasure of the past. Moreover, Pynchon suggests that because virtual space exists as a part of the capitalist system, it is prone to the same problems to which physical space (or meatspace, as hackers refer to it) is prone. As Pynchon sees it, it is unlikely that the internet can ever have a redemptive function for Americans. It can never provide meaningful transcendence even though it creates the illusion of it. It can, however, prompt Americans to rethink connectivity as a concept and develop nondigital human connections. Americans can come to make connections about history and conspiratorial mystery in productive ways, for instance in thinking about the simultaneous rise of the digital age and the age of terror. And they can come to develop meaningful familial and feminist relationships. These relationships can provide depth to human experiences and help resolve the sense of disconnection that defines the digital age and the age of terror. Ultimately, these relationships will allow Americans to buck systems of power and thus emerge, as Maxine does, as badasses in history: individuals who have the potential to create a better future that undercuts the hold that capitalist, technocratic systems sustain over Americans ("Is It O.K. to Be a Luddite?").

Like Pynchon, Jennifer Egan, the subject of this chapter's second section, has emerged as a literary chronicler of technology's history in the United States. As she explains in "Rewiring the Real," a 2012 conversation moderated by Willing Davidson at Columbia University's Institute for Religion, Culture, and Public Life, living "in New York for a long time" gives her the sense that 9/11 created a major "transformation" in history. But, she continues, "the big change that I witnessed in my lifetime is technological." For Egan, "the acceleration of technology," which is evocative of Wajcman's considerations of the accelerative feel of technology,[1] "feels like kind of the big story" that has created a division in her life—a sense of things as existing "before and after." "When I think about being in my early twenties," she elaborates, "partly what I think about, or maybe what I notice, is that [there exists] a certain kind of communication that I have all the time now that I didn't have then." In other words, connection in its new digital form distinguishes the twentieth century from the twenty-first for Egan, even though she suggests that there exists a "real contradiction" between her "relationship to technology as a consumer" and her "interest in it as a writer." In her personal life, she identifies as a "late adopter" of technology and confesses that her children view her as a "crazy crusader" who is trying to ban digital devices from their lives. They essentially see her as a modern-day descendant of the Luddites that Pynchon historicizes in "Is It O.K. to Be a Luddite?" By contrast, in her professional life, she fixates on digital culture. For instance, early in her career, she comments on the way that digitization influences the conceptualization and development of identity in her novel *Look at Me*. She portrays New York City–based model Charlotte Swenson as broadcasting to an online audience the metaphorical reconstruction of her identity following a life-changing car crash and traumatizing physical facial reconstruction with eighty titanium screws. More recently, in "Black Box," a work of fiction originally published as a series of tweets in 2012 on the *New Yorker*'s Twitter feed, Egan explores the simultaneously debasing and augmenting effects of digital technology on the female body. She tells the futuristic story of the undercover antiterrorist mission of a cyborg "citizen agent," presumably Lulu Peale, a character who first appears in *A Visit from the Goon Squad*.

My analysis of *A Visit from the Goon Squad*, arguably Egan's most noteworthy meditation on digitization, considers the relationship between history, art objects that both represent and shape history, and the digital world. In the thirteen interrelated chapters that comprise the chronologically nonlinear work, which Cowart has compared thematically to Wallace Stevens's "Thirteen Ways of Looking at a Blackbird,"[2] Egan predominantly tells the stories of Bennie Salazar, the nostalgic former punk-rocker-turned-music-producer who founded Sow's Ear Records, and Sasha Grady, Bennie's kleptomaniacal assistant. But Egan also creates a virtual web that is evocative of the World Wide Web and the web of connections that Pynchon portrays in *Bleeding Edge*. In Egan's book, this web exists among numerous characters who appear across time and space in scenes loosely or directly associated with the pre- and post-digital-age music industry in the United States. According to Aaron DeRosa, the book intimates that a "nostalgia for the future" exists in America in the aftermath of the 9/11 terrorist attacks that haunt the text, most notably via the rock concert at Ground Zero at the end of the book (89). There exists, to reference DeRosa's words, a longing for "a future exceptional status that would now never come to be" (89). And, as I argue, there exists, too, a nostalgia for the texture of pre-digital-age art and for pre-digital-age history. Although Bennie's nostalgia manifests predominantly as a condemnation of digital music and a valorization of its analog predecessor as a form of art, Egan opts against celebrating the past, anachronisms, or antitechnological ways of thinking and being unequivocally as Bennie does. Instead, she proposes that problems exist with purists of different kinds because they valorize the past or embrace the digital future without thinking critically about notions of progress. To move beyond the unproductive metaphorical binary code of purisms toward a productive future, Egan suggests embracing the possibilities that hybridity and disorder afford. She invites her readers to puzzle over her own book as a hybrid and political work of art that comes into dynamic interplay with a range of genres and media, for instance PowerPoints, vinyl records, and text messages. And she invites them to find redemption in ways that her characters cannot: by connecting their reflections on hybrid aesthetic objects such as her book with actions they take that function in ways analogous to the countercorporate actions of Pynchon's badasses. As Egan sees it, her readers' actions can ultimately shape twenty-first-century, digital-age history as a hybrid and ethical B-side to the American Century in the narrative of the United States that they create.

9/11, the Internet, and the Future of American History in Thomas Pynchon's *Bleeding Edge*

In *Bleeding Edge* (2013), the focus of the first part of this chapter, Pynchon exposes and probes the ways in which the circumstances that produce local and global history relate to the ones that produce digital technology as it exists in the digitally mediated moment from which he writes. He tells the story of the United States from March 2001 until mid-2002—just pre- and post-9/11—by way of telling the story of Maxine Tarnow, a Jewish, Manhattan-based fraud examiner who loses her license but continues with private investigations and who mothers two sons, Ziggy and Otis, from her deteriorating marriage to Horst Loeffler, a Lutheran, fourth-generation American from the Midwest. As Francisco Collado-Rodríguez suggests, Maxine functions as a twenty-first-century version of Oedipa Maas, the protagonist of *The Crying of Lot 49* who, as a child of the traditionalist 1950s, searches in the revolutionary 1960s "for the meaning of the United States of America" (229). Just as Oedipa investigates Trystero, an underground postal service that may or may not exist, Maxine investigates hashslingerz, a computer security firm run by Gabriel Ice, a Jewish billionaire who avoided financial devastation in the 2000 dot-com bust and who may have something to do with the death of Lester Traipse (the former owner of hwgaahwg.com who steals money from Ice) or even the attacks of 9/11, as suggested by funds he transfers to accounts in Dubai. Just as Oedipa's mind is bound in large part by the conservative 1950s in which she came of age, Maxine resembles numerous characters in Pynchon's 9/11 novel in that she is bound by history. At the beginning of the novel, Pynchon portrays Maxine as someone who "doesn't want to let go just yet" as she walks her sons, Ziggy and Otis, to the Otto Kugelblitz School on the Upper West Side (1). He portrays her as avoiding the fact that her sons no longer need her supervision. Hence he portrays her as representative of Americans who are attached to the past as he himself is attached to it. He is a historical novelist who likely feels nostalgia while critiquing it.

Pynchon showcases ways in which media culture sets the terms for the experience of nostalgia that characters and readers alike may have for the 1990s as the celebratory culmination of the twentieth century, which media mogul Henry Luce termed the American Century in his 1941 *Life* magazine editorial. According to the novel's narrator, memory has a "remote" as a television might (16). And by way of memory's remote, Maxine can hit "PAUSE, then STOP, then POWER OFF" as she recalls how she originally met amateur documentarian Reg Despard (16). Similarly, Pynchon's other characters and readers alike can—and do—flip through memories of history in the novel, especially the 1990s, a golden age in the United States because it saw the internet commercialize into a household staple and because Americans perhaps believed at that point in the idea of the end of history, to cite Francis Fukuyama's bold claim in *The End of History and the Last Man* that the spread of Western liberal democracy signified the end of the world's evolution. Lighthearted, passing references to Beanie Babies, Furby toys, *Johnny Mnemonic*, a "Tori Spelling marathon on Lifetime," and Ben & Jerry's chocolate peanut butter cookie dough–flavored ice cream (which got discontinued in 1997) pepper Pynchon's text (*Bleeding Edge* 94). They complement his overt references to nostalgia for appropriated Soviet culture and digital culture of the 1990s. For instance, Maxine encounters Driscoll Padget, a freelance web-page designer who has the Rachel haircut that Jennifer Aniston made famous in the 1990s on the TV series *Friends* and who is holed up in the abandoned hwgaahwgh.com listening to the bouncy Russian folk song "Korobushka," the "anthem of nineties workplace fecklessness" (43). It emanates from her 1990s Nintendo Gameboy as she plays Tetris while stealing bandwidth just before heading out to a bar that still serves Zima, a dated, Americanized, and watered-down appropriation of Russian vodka that Coors Brewing Company marketed and that adopts the Russian word for winter as its name. And all these references and moments in Pynchon's text build toward the most overt act of 1990s nostalgia in the novel: the 1999-themed party Gabriel Ice throws that emphasizes "instant nostalgia" (301). It puts front and center the reality of the Cold War victory celebration that comes to coexist with Y2K anxiety and Americans' sustained romance with the apocalypse that the Cold War promised (301). It spotlights the peculiar mentality and values that 1990s Americans had and wish they could retain.

In turn, American ways of remembering the Cold War transform because the 1990s, which exist as a victorious and hence celebratory moment in American history, mediate American memory. Throughout *Bleeding Edge*, Pynchon romanticizes a Cold War that was anything but romantic for historical Americans who lived through it in fear of nuclear war, practicing futile duck-and-cover exercises as schoolchildren. Much as Shteyngart creates a web of intertextual relationships with Slavic literary works in *Super Sad True Love Story*, Pynchon invites his readers to unearth a nondigital web of Soviet-related phenomena in the puzzling matrix that his novel constitutes: the "matrix from which [history] springs," to reference Cowart's words ("'Down on the Barroom Floor'"). For instance, Pynchon gives Maxine a Slavic surname, he makes mention of the fact that Natalie Wood was "born Natalia Nikolaevna Zakharenko" (a typical Ukrainian name), and he observes that Reg and Maxine meet at what is likely Veselka, a "24-hour Ukrainian joint in the East Village," a historically Ukrainian part of Lower Manhattan that speaks to influxes of immigrants who arrived predominantly during or after World Wars I and II (*Bleeding Edge* 27, 90). More overtly, Pynchon positions Russian mafioso Igor Dashkov and Misha and Grisha, recent graduates of a school for hackers in Moscow, as lovable comic relief in the novel. Igor eats Russian ice cream that has high butterfat content, and Maxine sees his act as evidence of "Soviet-era nostalgia, basically" (162). And he outright confesses to feeling "[s]imple nostalgia" for the Cold War when he asks Maxine to watch Reg's digital video footage of the apparent Stinger missiles on the rooftop of the Deseret luxury apartment building on the eve of 9/11 (273). Similarly, Misha and Grisha have a comic function in Pynchon's text as they engage in a parody of real American terror as post-9/11 characters and real-life Americans know it. Their adventurous trip to upstate New York to set off a pulse to disable electrical power to Gabriel Ice's server farm in Lake Heatsink might be read as an act of cyberterrorism, but instead, the "LESTER TRAIPSE MEMORIAL PULSE [...] barely gets onto the local news upstate, forget Canadian coverage or the national wire, before being dropped into media oblivion" (468). At best, it enables readers to experience nostalgia for a simpler era that had a knowable Russian enemy, not a largely unknown fanatical Muslim terrorist one.

By showing ways in which his characters and readers may long for, chuckle about, and romanticize Cold War history, Pynchon suggests that history in the digital age moves not necessarily in entropic ways toward disorder, as he intimates in much of his oeuvre, beginning with "Entropy" as a short-story meditation on the scientific concept. Instead, it moves through unsettling cycles. He implies that the 9/11 attacks change the United States, but, paradoxically, they changed it by allowing Americans to more fully revert back to pre-1989 modes of thinking and being despite the clear trauma that ensues after 9/11. Thanks to 9/11, America replaces a Soviet enemy with an Islamist one that puts everyone "on edge" (316). As March Kelleher, Ice's mother-in-law, puts it in her blog, after 9/11 "a hole quietly opened up in American history, a vacuum of accountability, into which assets human and financial begin to vanish. Back in the days of hippie simplicity, people liked to blame 'the CIA' or 'a secret rogue operation.' But this is a new enemy, unnamable, locatable on no organization chart or budget line—who knows, maybe even the CIA's scared of them" (399). And thanks to 9/11, American feelings about the past emerge as no longer nostalgic because fear of terrorism in the present that is reminiscent of Cold War anxiety renders nostalgia moot. To put my point another way, fanatical Islamist terrorists fill the void, or are constructed to fill the void, that the Soviet Union and the Cold War leave behind, and evidence of this newfound and noteworthy absence of nostalgia and of regression exists, for example, in the narrator's remark that Maxine's neighborhood "reverts to its usual insufferable self" by Thanksgiving (380). It exists in Driscoll shaving her head and getting a Rachel-haircut wig to save money—and then eventually "[m]oving on" from the haircut altogether (366). It likewise exists in rhetorical and historical phenomena that come to characterize the post-9/11 world. As Pynchon's narrator points out, the term "Ground Zero," which describes the former site of the World Trade Center, historically exists as a "Cold War term taken from scenarios of nuclear war so popular in the early sixties," and as a result, it gets "people cranked up in a certain way" (328). Furthermore, as Maxine learns in a history lesson from her politically liberal father, Ernie, after she makes the faulty claim that "[n]obody's in control of the Internet," which exists as a staple of American life by the start of the twenty-first century, the U.S. government developed the internet to enable communication in the case of a nuclear attack (419). In Ernie's words, "Your Internet, back then the Defense Department called it DARPAnet, the real original purpose was to assure survival of U.S. command and control after a nuclear exchange with the Soviets"—a kind of survival to which all Americans in the age of terror are now privy (419).

Pynchon complements his discussion of the problem of satisfying nostalgia by repeating history with allusions to the historical nondissemination of history via extant and increasingly digital media platforms or the outright erasure of history, phenomena that he sees as equally unsettling. Most notably, he gestures toward the intersection of the corrupt history behind the romanticized history of the Cold War and the history behind September 11. As the novel's narrator explains, Nicholas Windust, a previous and possibly current federal operative who is perhaps Traipse's killer and who has an affair with Maxine, has his first field operation "as an entry-level gofer" in Santiago, Chile, on a lesser-known 9/11 (108). He sees covert action among homegrown "neoliberal terrorists" on 11 September 1973, when the

U.S. government enacted a sanctioned act of terror against Chile during the Cold War by backing Chilean forces that killed the democratically elected Marxist president, Salvador Allende, to stop the spread of anti-American, anticapitalist communism (108). The kind of history that 11 September 1973 represents is an unsanctioned and hotly debated history because of who has the power to write history and who lacks that power. Chilean supporters of Allende certainly saw and likely continue to see the United States as having engaged in condemnable if not outright criminal or terrorist activity. But according to U.S. nationalist rhetoric that Americans absorb through the mass media, American efforts in Chile existed and continue to exist in memory as liberatory. And because, on a global scale, American capitalists have more power to write history than Chilean Marxists have, the truth of what happened on 11 September 1973 remains obscured and hard to understand, not unlike the labyrinth that constitutes a Pynchon novel rife with conspiracy. Indeed, the obscurity of the first 9/11 speaks to what Pynchon sees as the inevitable obscurity of the second one. Although traumatized Americans receive a nationalist narrative through the media that speaks of clear heroes, victims, and villains, there exists something conspiratorial, deceptive, and reductive about this nationalist narrative. As Maxine's friend and Horst's former mistress, Heidi, explains in a more critical reading of history, "11 September infantilized this country. It had a chance to grow up, instead it chose to default back to childhood" (336). And to appropriate the narrator's words about Maxine and Shawn, the United States finds itself "down here on the barroom floor of history" (339). To further quote the narrator's words, Americans in this infantilized, devastated nation may feel that "the country itself may not be there anymore" because it is "being silently replaced screen by screen with something else" (339).

The narrator's attention to the way in which screen culture produces reality and thus history speaks to *Bleeding Edge*'s distinction from *The Crying of Lot 49* as a quest narrative. More importantly, it speaks to the influence of new media technology on approaches to archiving and searching the archives of history, a subject that David Haeselin explores in "Welcome to the Indexed World: Thomas Pynchon's *Bleeding Edge* and the Things Search Engines Will Not Find." In talking about the laughable restaurant Muffins and Unicorns, where someone "has been using phantomware to falsify cash-register receipts," Felix Boïngueaux may observe that "[w]e're beyond good and evil here" and that technology is "neutral," but Pynchon disagrees (Pynchon, *Bleeding Edge* 87, 89, 89). For Pynchon, as for Don DeLillo in *Cosmopolis* and Dave Eggers in *The Circle*, technology is always already implicated in the American capitalist system. As Pynchon suggests, new media technology builds on old media technology that shapes memory as a thing to be controlled by a remote, at least once characters such as Maxine and readers process its implications. To reference Mitchum Huehls's words, by the twenty-first century, conspiracy is "shaped like a database," and hence quests and searches of all kinds involve not only physical reality and the human mind that reads physical reality, but cyberspace and the bank of digital memory that it contains (869). Evidence of the new shape of searching the annals of history exists in Maxine's reliance on digital information. She attempts to make sense of what she sees on a mysterious DVD purportedly sent to her by Reg and on a flash drive that a bicycle deliveryman drops off for her. Evidence of this new way of searching also exists in the digital banks of information that contain data from surveillance akin to that of Eggers's *The Circle*. As Windust tells Maxine, he knows about her online movements because digital devices know about them. In his words, "People at my shop have learned of your interest, we assume professional, in the finances of hashslingerz.com" because, to borrow Windust's words, "[n]o keystroke" gets "left behind," a phrase evocative of conservative educational policy and evangelical Christian notions of the apocalypse (Pynchon, *Bleeding Edge* 105). The powers that be, be they corporations or the U.S. government, likewise track movements in the physical world through digital means and store the data they accrue in digital archives. As Traipse explains before his apparent suicide and possible murder, he is "[s]hitcanning" his cell phone because he thinks "there's a tracking chip on it"—and his assumption is likely right (173).

Pynchon's ideas seem to dovetail with those of Michael Patrick Lynch, who, in *The Internet of Us*, explores how information technology is in disconcerting ways easing "the burden of remembering" and thereby "affecting what we know and how we know it" (xvi, 6). Specifically, Pynchon suggests that digital memory threatens to change human ways of thinking and human desires to know, chronicle, and understand history. His readers see the effects of digital media on humanity in relation to history in his representation of DeepArcher, a bleeding-edge application that is still in beta form and that is designed by Justin and Lucas, two Silicon Valley transplants to Silicon Alley, located in the Flatiron District of Manhattan. When Maxine first enters DeepArcher, it confounds her sense of time as a building block of history and shows the peculiar connection that exists between time and technology—a connection that Dr. Zoot's time machine in *Against the Day* also spotlights. She asks, when she reemerges from its virtual depths, "What time is it?" (Pynchon, *Bleeding Edge* 77). Furthermore, Pynchon suggests that DeepArcher makes desirable for the masses—not only for those who have the power to write or not write history—the erasure of history. As Vyrva McElmo, Justin's wife, tells Maxine, everyone is after the DeepArcher source code—even Gabriel Ice, who seeks to hide hashslingerz's transaction history—because of its ingenious "security design" (36). Thus, DeepArcher realizes the paronomastic metaphor of its name by allowing its users to engage in a departure not only from the reality of meatspace but from the mass surveillance of the surface web in the deep web, a part of the World Wide Web that search engines do not index and that contains the dark web, where extensive illegal activity occurs. As Justin puts it, DeepArcher "forgets where it's been, immediately, forever" (78). It produces no history of movements in digital space, and, for better or worse, it helps render humanity as free of history.

Just as post-9/11 ways of thinking and being find their counterpart in familiar but equally anxious Cold War–era manners of thinking and being, cyberspace that aims to be history-free finds its counterpart in meatspace, most notably of the gentrified variety that covers over history and that Pynchon critiques throughout the novel. In *Bleeding Edge*, Pynchon shows how power comes with ownership of space, be it physical or virtual, and even though virtual space seems new, vast, and free of limitations and problems that exist in meatspace. Even though virtual space is space that Eric Schmidt and Jared Cohen in *The New Digital Age: Transforming Nations, Businesses, and Our Lives* characterize as an unregulated "experiment involving anarchy," it relies on meatspace in important ways (3). As Andrew Blum suggests in *Tubes: A Journey to the Center of the Internet*, the internet cannot exist without physical cables and real-life spaces that house servers which store data. And Ice, known as a "bandwidth hog," focuses his efforts on acquiring physical space and material that allow his digital space to thrive, for instance Lake Heatsink, but also real-world fiber: cable that he can pull "through conduit" or "hang" or "bury" or "splice" (Pynchon, *Bleeding Edge* 156, 465, 465, 465, 465). In other words, he colonizes meatspace in order to colonize cyberspace. And his troubling capitalist and colonial efforts resemble acts of gentrification that pepper Pynchon's text. For Pynchon, the death of neighborhoods, be they physical or virtual, signifies the death of history. As the mournful narrator puts it in describing March's dying Upper West Side Manhattan neighborhood, "Someday very soon this will all be midtown, as one by one the sorrowful dark brickwork, the Section 8 housing, the old miniature apartment buildings with fancy Anglo names and classical columns flanking their narrow stoops, and arch-shaped window openings and elaborate wrought-iron fire escapes rapidly going to rust, are demolished and bulldozed into the landfill of failing memory" (267).

And just as March's neighborhood is dying, so too is the dream of the internet that existed in the 1990s—the dream of the internet as a free space that Deep-Archer in many ways represents. Certainly, Jaron Lanier speaks to this problem in *You Are Not a Gadget* when he observes that "[s]omething started to go wrong with the digital revolution around the turn of the twenty-first century. The World Wide Web was flooded by a torrent of petty designs sometimes called web 2.0. This ideology promotes radical freedom on the surface of the web, but that freedom, ironically, is more for machines than people" (3). And after 9/11 in Pynchon's novel—so around the same time—DeepArcher goes open source, thereby realizing IT expert and foot fetishist Eric Outfield's prediction that "colonizers are coming" and that "[l]ink by link, they'll bring it all under control, safe and respectable" (241).

Pynchon suggests that what had once been a deep, digital refuge from mainstream American capitalist culture—what once had potential to create some semblance of meaningfully connected experience despite its ahistorical nature—gets flattened out. And in its flattening, the internet ultimately cannot have any real redemptive function in American history. DeepArcher provides disconnection as opposed to connection as it emerges as just another part of the mainstream flat world when it welcomes "in half the planet, none of them who they say they are" (426). And the internet more generally loses any redemptive capability that its users may have believed it to have. Although Maxine sees visions of the dead while in DeepArcher, including "likenesses" of the victims of 9/11 and visions of both Traipse and Windust, these visions exist as pale reflections of historical and material reality at best (357). At worst, these visions exist as evidence of digital identity theft, which Joshua Ferris portrays in *To Rise Again at a Decent Hour*. As the narrator puts it after Maxine runs into a virtual Windust, "So who was she talking to, back there in the DeepArcher oasis? If Windust, judging by the smell, was already long dead by then, it gives her a couple of problematic choices—either he was speaking to her from the other side or it was an imposter" such as Gabriel Ice or a "random twelve-year-old in California" (411). If there exists, in or on the internet, an afterlife, as Foer suggests in *Here I Am* and as Ice intimates when he tells his wife, Tallis, that there is "no scenario where I die," then it is certainly not heavenly (473). People become, at best, akin to items for sale online on eBay. They live in "the collectors' market" as a parody of an "afterlife" (435). They become pixels themselves and hence akin to the devastated World Trade Center, which terrorists "blew" to "pixels," to give rise to a potentially hellish new American condition in the new millennium (446).

Bleeding Edge, however, presents a relatively hopeful vision for American life amid the digital, capitalist, and governmental systems that work to flatten if not deaden American experience into meaninglessness, as evidenced by Pynchon's focus on depth as a theme and the possibilities that depth affords. Although DeepArcher as bleeding-edge technology in its pre-9/11 form and the deep web collapse into a capitalist system that Pynchon views as toxic, he posits that there may exist a redemptive function to engaging in efforts to make meaningful nondigital connections among apparently disparate pieces of information. And these efforts can lead to a deeper understanding of the self and of post-9/11 America. They might repair the metaphorical bleeding edge that the 9/11 attacks leave in their wake. Certainly, digital-age technology changes what it means to be connected. But it fails to suppress the joy of trying to make the nondigital-age connections that Maxine attempts to make and that readers of literature so appreciate. As a former certified fraud examiner, Maxine knows well how to "look for hidden patterns," and she follows breadcrumbs as Oedipa does in *The Crying of Lot 49* in a valiant but failed effort to unearth conspiracy (22). For instance, Maxine smells the scent of the 9:30 Club at the site of Traipse's apparent suicide—the same scent she smells when she is in Windust's presence—and she imagines a connection, but she never confirms that Windust is Traipse's killer. Likewise, she sees that a connection may exist between the DVD footage she receives of the Stinger missiles on the rooftop of the Deseret and Traipse's death in the Deseret—and perhaps even with the 9/11 terrorist attacks—but she never fully sees a clear picture of how these disparate elements connect. If revelation constitutes an understanding of the connections that come together to comprise conspiracy, then much as revelation trembles "just past the threshold" of Oedipa's understanding in *The Crying of Lot 49*, it trembles past Maxine's understanding in *Bleeding Edge* (*The Crying of Lot 49* 14). Indeed, when the Rudy Giuliani–sponsored construction that symbolizes the gentrification of old New York wakes Maxine from her dream, it should perhaps signal to her the significance of the problem of gentrification to physical and virtual spaces as containers of history. Instead, she wakes up unable to see the connections she is on the verge of making. According to the narrator, she feels that "any message is corrupted, fragmented, lost" (*Bleeding Edge* 210).

Although Pynchon creates a relatively parallel experience for his protagonist and his readers in *The Crying of Lot 49* because revelation in the form of understanding of conspiracy escapes both protagonist and reader, he creates a distinction between the experience of his protagonist and that of his readers in *Bleeding Edge*. Readers, like Maxine, seek to make nondigital connections that lead to deeper understanding. But unlike Maxine, readers are equipped with historical knowledge that characters have yet to know. They know literary history, for instance *The Crying of Lot 49*, and thus they know that by having Maxine mention that her "Tupperware party" is on Tuesday, Pynchon is referencing the opening line of *The Crying of Lot 49*, which portrays Oedipa as coming home from "a Tupperware party whose hostess had put perhaps too much kirsch in the fondue" (*Bleeding Edge* 220; *The Crying of Lot 49* 1). More importantly, readers know that the internet continues to thrive in the wake of the dot-com bubble bursting, and they know that the 9/11 attacks are on the horizon in Pynchon's book. In other words, although they may not understand connections that may constitute a conspiracy, they know a non-conspiracy-related revelation that awaits Pynchon's characters as pre-9/11 Americans: the 9/11 attacks as a revelation of history. They know from the novel's start what Shawn says to Maxine after 9/11 transpires in the world of the novel: that 9/11 was a moment of revelation that showed Americans "exactly what we've become, what we've been all the time" and not a moment that changed everything or anything (*Bleeding Edge* 340). As a result, readers read information that appears in the novel in radically different ways than Maxine or other characters read it. This information includes, for instance, Shawn's anti-Muslim sentiments or March's remarks at the Otto Kugelblitz School about the "Bush family" doing "business with Saudi Arabian terrorists" (53). Readers likewise have emotional experiences because of their knowledge of future American history—emotional experiences that Maxine and other characters cannot have because they lack knowledge. For example, they feel the emotional reality of 9/11 when Sidney Keheller, March's ex-husband, drives March and Maxine past a view of the World Trade Center "leaning, looming brilliantly curtained in light" or when Horst, Otis, and Ziggy describe playing violent video games that involve a "postapocalyptic New York" with the on-screen World Trade Center "leaning at a dangerous angle" (165, 292, 292). To borrow Pynchon's narrator's words, readers see that after 9/11, "[t]here's no way not to talk about the Trade Center"—even if the subject of conversation is pre-9/11 American life or fiction about the pre-9/11 United States (324).

Even though no conspiracy emerges in *Bleeding Edge*, a stand-in for conspiracy develops in the form of 9/11 as a deeply momentous event because of the way in which it connects everything and everyone in America. Specifically, the 9/11 attacks foster a connection between the twentieth century and the twenty-first. They exist as apocalyptic and fulfill the desire for the apocalypse that 1990s Americans had at the moment at which the historical narrative of the twentieth century came to a relatively anticlimactic end. The 9/11 attacks also connect readers of history and readers of fiction with one another. They conjoin historical reality as it comes to exist in the annals of history with imagined expectations for the end of the second millennium that readers of fictional narratives may have because, as Frank Kermode argues in *The Sense of an Ending*, fiction—especially fiction that resists postmodern tendencies toward disorientation through nonlinear narrative—mimics the structure of the Christian Bible, which ends in Revelation and thus in a dramatic ending that satisfies readers' "hunger for ends and for crises" (55). Hence Pynchon's representation of 9/11 as the apocalypse in *Bleeding Edge* as fiction also connects *Bleeding Edge* as an arguable work of postmodern fiction with post-postmodern fiction, whatever form it might take. It connects postmodern fiction with unironic fiction, a notion that Heidi acknowledges when she argues in her article for the fictionalized *Journal of Memespace Cartography* that "irony, assumed to be a key element of urban gay humor and popular through the nineties, has now become another collateral casualty of 11 September because somehow it did not keep the tragedy from happening" (Pynchon, *Bleeding Edge* 335). It connects *Bleeding Edge* with the notion that postmodernism as an aesthetic movement involves more than mere irony because it shows that a postmodern novel can tackle a serious subject such as 9/11 in a meaningful way and produce thought-provoking results and a way of thinking that Pynchon views as valuable.

By inviting readers to consider the connections that 9/11 cultivates, Pynchon in turn invites them to contemplate the connection—or lack thereof—between terrorism and digital culture, both of which penetrate deep into twenty-first-century American consciousness and both of which as a result function as the key subjects of his novel. Terrorism and digital culture distinguish the twenty-first century from the twentieth in the narrative of American history that Pynchon attempts to illuminate—a narrative that speaks in ways to Hamid's *Exit West*. Fundamentalist terrorism, an apparently antimodern phenomenon because fanatical Islamist terrorists scorn modernity, exists as a response to modern manifestations of Western capitalist wealth: manifestations such as digital culture, a quintessentially modern phenomenon. Yet there exists something flat and digital about terrorism given that Americans for the most part viewed and learned about 9/11 as an act of terror by way of screens, be they television screens or digital screens that gave them access to the internet. And there exists a terroristic quality in digital culture, as best evidenced in the novel by the first-person shooter game designed by Justin and Lucas—the game that Maxine's sons and Vyrva's daughter Fiona play near the beginning of the novel. As Pynchon portrays it, the first-person shooter game allows players—*children*—to desensitize themselves to mass death on a deep, subconscious level. They can and do disable the "splatter options" when they play (33). As a result, to appropriate Vyrva's sarcastic words, they render acts of terror as mom approved (35). Moreover, digital first-person shooter games as Pynchon represents them morph Americans' conceptions of reality. As the narrator puts it, the squatters that Maxine and Windust encounter while walking down "[d]esolate corridors, unswept and underlit," are "like targets in a first-person shooter" (258). The reality of their human existence gets flattened in significance until it emerges as equal to an insignificant fictional bystander in a video game. And when Maxine and Windust have sex soon after passing by the squatters, digital culture pervades their sexual encounter. It flattens it out and cultivates violence in it as well. According to the narrator, Windust's "hands, murderer's hands," grip "her forcefully by the hips, exactly where it matters, exactly where some demonic set of nerve receptors she has been till now only semi-aware of have waited to be found and used like buttons on a game controller" (258).

The ways in which ubiquitous digital culture promotes increasingly prevalent terrorist activity intimates that twenty-first-century humanity is changing in profound and detrimental ways, and Pynchon intimates that sustaining deep human connections exists as a countercultural if not outright rebellious act that might counter the flattening and terroristic effects of digital-age American life. For Pynchon, sustaining deep human connections somewhat paradoxically renders individuals into contemporary badasses of the sort that he describes in "Is It O.K. to Be a Luddite?" As Pynchon explains in the essay, published in 1984 in the *New York Times Book Review*, Ned "King" Ludd, a possibly fictional figure who is characterized as the first Luddite of history, was no technophobe, even though the term *Luddite* signals technophobia in the contemporary world. He functions as a "dedicated Badass" because he destroys knitting machines out of anger for the way in which the machines were "putting people out of work." In other words, he functions as a badass because of the way his actions aim to buck the industrial system as the dominant system of power in the nineteenth century. As Pynchon explains in the essay, "There is a long folk history of this figure, the Badass. He is usually male, and while sometimes earning the quizzical tolerance of women, is almost universally admired by men for two basic virtues: he is Bad, and he is Big. Bad meaning not morally evil, necessarily, more like able to work mischief on a large scale."

But by the twenty-first century, Pynchon revises his vision of what constitutes a badass, as evidenced by his representation of Maxine, the character whom I read as the clear badass of *Bleeding Edge*.[3] Not male or big or bad in any conventional sense of the term, she functions as a badass because she bucks the dominant system of the postindustrial twenty-first century: the male-dominated world of digital technology that disseminates feelinglessness in an effort to dissipate human feeling and human, nondigital connections. She bucks the patriarchy as Kristin Roupenian aims to buck it through the publication of "Cat Person," a short story that vilifies not men in general but the ways in which many men treat women in the United States and in the world.

In *Bleeding Edge*, Pynchon celebrates the badass nature of familial connections, which he establishes in his own life by the time of the novel's publication by marrying Melanie Jackson, his literary agent, and raising with her their son, who was roughly Otis's age when the 9/11 terrorist attacks occurred. In particular, by way of portraying a near-broken married couple that redevelops some semblance of a connection with each other in *Bleeding Edge* (a connection that stands in stark contrast to the marital disconnection that Foer portrays in *Here I Am*), Pynchon suggests that sustaining familial relationships can produce a depth of character in twenty-first-century America that can trump the kind of "high-powered connections" that Ice maintains through the capitalist system he venerates (137). Although Pynchon opens the novel with a tacit proclamation of Maxine's familial disconnection—the fact that some systems still list Maxine not as Maxine Tarnow, her maiden name, but as Maxine Loeffler, her married name—the novel showcases the start of a possible reconnection with Horst, who sustains love for Maxine, or at least for the security that their marriage affords in the insecure, post-9/11 times. Near the novel's end, Maxine and Horst find themselves sleeping together in a literal (as opposed to a euphemistic) sense. As the narrator explains, Maxine and Horst "may have dozed off on the couch for a second," as evidenced by the time lapse and the sudden fact of daybreak (467). And the incident suggests an intense intimacy that never manifests between Maxine and either of the other lovers she takes over the course of the novel. By the novel's final pages, intimate domestic questions that involve Horst occupy Maxine's thoughts. As the narrator explains, "she begins to wonder, actually, whose turn it is to take the kids to school" (476). The children create an important connection between Maxine and Horst, and they may save their marriage.

Pynchon likewise celebrates as badass the deep connections that women—especially mothers—sustain in the male-dominated world that the twenty-first century constitutes because of the prevalence of capitalism, digital technology, and terrorism, all of which function as domains governed predominantly by men. Near the end of *Bleeding Edge*, Maxine engages in an act of solidarity that is evocative of her history of opposing through protest "the co-opping frenzy of ten or fifteen years ago" with March (54). She does not work mischief on a large scale per se, because only corrupt capitalist or digital systems can work on such a scale by the twenty-first century. But she works with March and Tallis, March's daughter and Ice's soon-to-be ex-wife, to liberate Tallis from Ice's control. After Tallis and Ice argue in the street outside of March's home, Maxine drives Horst's 1959 Impala, a quintessential muscle car definitive of masculinity for American men, to go off grid, where Ice's "[k]iller drones" will be unable to find them (474). Hence she functions as a modern-day badass because she so effectively uses a man's tool to hide from Ice as a man's man of the twenty-first century—a man who has his hand in capitalism, digital technology, and perhaps even terrorism. As she explains to her passengers, "we're going to keep off bridges, out of tunnels, stay right here in town, and go hide in plain sight" (474). And in so doing, she changes the physical world. She morphs a man's muscle car into a space for women's safety and dreams of the future, a womb out of which Tallis can be reborn as a liberated woman. As the narrator explains, the forty-year-old "Luxury Lounge interior" with "not-yet-damped vibrations of Midwest teen fantasies that've worked their way into the grain of the metallic turquoise vinyl, the loop-carpet floor mats, the ashtrays overflowing with ancient cigarette butts, some with lipstick shades not sold for years, each with a history of some romantic vigil" has "wrapped them, brought them in from the unprofitable drill-fields of worry about the future, here inside, to repose, to unfurrowing, each eventually to her own dreams" (475). And "[n]ext thing anybody knows, it's morning," and they wake to find themselves safe from Ice because of their solidarity with one another—a feminist solidarity that eludes Oedipa as a relatively solitary woman in the man's world in which she must operate in *The Crying of Lot 49* (*Bleeding Edge* 475).

Ultimately, Pynchon posits that potential for a deeply interconnected, socially just, and ethical future requires not only interconnection and solidarity among women but a meaningful connection with the generation of children that the Maxines and Horsts, the Vyrvas and Justins, the Tallises and Gabriels, and, of course, the Melanie Jacksons and the Thomas Pynchons have produced. Notably, at the heart of *Bleeding Edge* is the perplexing vision of an apparent child that Maxine encounters when she sneaks into a tunnel beneath Ice's Montauk, New York, property—a vision that showcases a perversion of childhood in that the child by stature in it is anything but childlike. Upon descending deep into "the terminal moraine," Maxine sees a space of "Cold War salvation down there, carefully situated at this American dead end" at the end of Long Island, where Pynchon spent his own childhood (194). She sees what looks to be a child "in a child-size fatigue uniform, approaching her now with wary and lethal grace, rising as if on wings, its eyes too visible in the gloom, too pale, almost white" (194). As a result, she runs in fear. Pynchon never explains whether the child is real. It could be Kennedy Ice, Gabriel and Tallis's son, whom Maxine sees. Or perhaps she sees an unsettling product of Ice's Montauk Project, an underground weapon-making operation that functions as a parody of the Manhattan Project, which produced the atomic bomb during World War II to usher in the start of the Cold War that *Bleeding Edge* in many ways commemorates and critiques. Alternately still, Maxine perhaps sees a projection of her subconscious mind, the nightmare reality that she fears will manifest for her children if the twenty-first century rolls on in accord with the militaristic and capitalistic values of the twentieth. Pynchon never specifies.

But he does specify that Maxine as the novel's heroine showcases heroism by sustaining a connection to her children as they grow up, as evidenced by her actions at the novel's conclusion. Although the novel comes full circle to complete a cycle of history, ending where it began (with the domestic scenario of Maxine's children walking to school), Maxine shows evidence of growth. Instead of walking with her children who no longer need a chaperone, as she does at the beginning of the novel, Maxine trusts her children to walk to school alone. And she puts faith in her children to manage the walk perhaps because she has seen and come to respect the kinds of choices they make in the digital world. She has seen the real world that they might be capable of building in real historical time because she has seen the impressive post-9/11 world that they produced in the deep web—a digital New York City rooted in a respect for history that exists for the people with the "old Hayden Planetarium, the pre-Trump Commodore Hotel, upper-Broadway cafeterias that have not existed for years, smorgasbords and bars offering free lunches, where regulars hang around the door to the kitchen so they can get first shot at whatever's being carried in" (428). This is a New York City that unfortunately "can never be" because real-world time cannot run in reverse to any productive end (428). But it is a New York City created by exemplary members of a generation who can inform emergent ethics in the real New York and in the real world. Indeed, the generation of children that Otis and Ziggy represent has potential to carry on the tradition of gloriously badass behavior thanks to the mothering of Maxine. It is a generation of children that can buck capitalist, digital, and terrorist systems in emergent history as Maxine has. Thus, it is a generation that has potential to create a more ethical future beyond the present limits of the digital age and the age of terror.

Digital Art, American Nostalgia, and Redemption through Hybridity in Jennifer Egan's *A Visit from the Goon Squad*

Whereas Pynchon centers *Bleeding Edge* predominantly on the intersection of digitization and history, Egan focuses *A Visit from the Goon Squad* (2010) on digitization and the history of aesthetics. In the novel, she tells the story of the American music industry's transition from analog music played on vinyl records or magnetized tape to digital music played on compact discs or digital devices such as iPods and smartphones. The aesthetic history that she presents, as she sees it, is complicated. As Egan portrays it through Bennie's perspective and as many real-life music critics agree, the digital age has flattened music in more ways than one, to echo Friedman's language from *The World Is Flat*. According to Katrina Morgan, "Sound waves contain an infinite number of points," but computers "cannot store infinite amounts of information." Thus, analog music "captures a physical process," but digital music "uses mathematics to reduce the process to finite bits of information" (Morgan). As Egan's narrator articulates this concept, in listening to digital music, Bennie "listened for muddiness, the sense of actual musicians playing actual instruments in an actual room. Nowadays the quality (if it existed at all) was usually an effect of analog signaling rather than bona fide tape—everything was an effect in the bloodless constructions Bennie and his peers were churning out" (*A Visit* 22). But, as the narrator continues, Bennie knows that the corporate music he produces to "satisfy the multinational crude-oil extractors he'd sold his label to five years ago" is "shit" (23). According to the narrator, it is "[t]oo clear, too clean. The problem was precision, perfection; the problem was *digitization*, which sucked the life out of everything that got smeared through its microscopic mesh. Film, photography, music: dead. *An aesthetic holocaust!*" (23).

Just as the United States witnesses a transformation in music that Bennie laments, the country likewise witnesses a broader transformation and flattening as the twentieth century rolls toward the twenty-first. It witnesses dramatic changes in America that take on the feel of changes in history, and they surprise characters in Egan's text because her characters live in the past, as do Pynchon's characters in *Bleeding Edge*. For instance, New York City publicist Dolly (La Doll) Peale observes that in attempting to throw a lavish New Year's party for A-listers that is evocative in its cultural stature of such historically significant U.S. events as Woodstock, she had "overlooked a seismic shift" in America (143). The event she conceptualizes and puts on resembles Gabriel Ice's party in *Bleeding Edge*. It crystalizes "an era that had already passed" as opposed to crystallizing the period in which she lives. Similarly, celebrity journalist Jules Jones describes a remarkable transformation he sees in the United States—and a flattening—after his release from a stint in prison for attacking actress Kitty Jackson, a vapid actress who is an associate of La Doll's. As Jules puts it in a description of the way in which prison traps him in the past, "I go away for a few years and the whole fucking world is upside down" (123). He continues by making reference to the interplay of the effects of digitization and the 9/11 terrorist attack as an attack on American capitalism that levels or flattens formerly tall symbols of it, a notion toward which Pynchon also gestures: "Buildings are missing. You get strip-searched every time you go to someone's office. Everybody sounds stoned, because they're e-mailing people the whole time they're talking to you" (123).

As Egan characterizes it, the primary driving forces in the dramatic technological and broader national transformations that she portrays resemble those that Pynchon outlines. They include twentieth-century American capitalism that seeks to retain U.S. prominence through technological development and other means and twenty-first-century Islamist terrorism as a violent response to American capitalism. In the twentieth century she portrays, Americans' interest in wealth, celebrity, and image supersedes a concern for ethical action, and capitalism's colonial logic functions as the threat to which al-Qaeda terrorists respond in the form of their attacks in 2001. As Jules describes American values and ethics near the end of the 1990s, "when you're a young movie star with blondish hair and a highly recognizable face from that recent movie whose grosses can only be explained by the conjecture that every person in America saw it at least twice, people treat you in a manner that is somewhat different—in fact that is entirely different—from the way they treat, say, a balding, stoop-shouldered, slightly eczematous guy approaching middle age" (168). And twentieth-century American capitalist values bleed into the twenty-first century. According to the narrator, what La Doll might be celebrating at her lavish party involves the fruits of capitalism for American elites: the fact that "Americans had never been richer, despite the turmoil roiling the world," an observation that is evocative of DeLillo's portrayal of the American elite in *Cosmopolis* (Egan, *A Visit* 141). Indeed, in portraying the perspective of Bennie's former bandmate Scotty Hausmann, the narrator observes that New York's "extravagance felt wasteful" (103). And extravagance and wastefulness paired with questionable ethics pepper Egan's book, as perhaps best demonstrated by La Doll taking on a genocidal dictator, General B., as a client and promising to improve his public image via carefully staged and disseminated photographs that cover up his historical atrocities.

In portraying dramatic transformations in America and in focusing on the economic forces that fuel these transformations, Egan invites readers to reflect on different approaches to viewing history and time, the subject to which her epigraphs from Marcel Proust's *In Search of Lost Time* allude and the goon that her title references. As the former Conduits guitarist Bosco says to Stephanie, Bennie's first wife, and as Bennie reiterates to Scotty near the book's close, "Time's a goon, right?" (127, 332). Certainly, time establishes itself as a goon—as a thug who beats up on humanity—in part because of the ways in which characters cling to youth and the past. As music producer Lou Kline puts it in what readers likely view as a laughable state of denial, "I'll never get old"—even though he already is old when he utters the phrase and finds himself hospitalized after a second stroke within the pages of Egan's book (56). And like Lou, characters in Egan's text resist growing up as much as they resent growing old. Hence they fail to see the passage of time as potentially meaningful. Notably, Sasha at the beginning of the book and her ex-lover Alex by the book's end hear time not as melodic or profound but as potentially annoying and ever present. They hear time in this way because they fail to change or understand positive effects of the passage of time. They hear time as the aesthetic and functional opposite to nondigital music in Egan's book: as a sound that resembles the constant and subtle noise of a digital device. For Sasha, who passes time talking with Coz, her therapist, about her kleptomaniacal acts but fails to write "a story of redemption, of fresh beginnings and second chances," the sound of time exists as a "faint hum that was always there when she listened" (8–9, 18). And for Alex, who attempts but fails to relive a moment of his youth spent with Sasha by visiting her old apartment with Bennie's help, it exists as an unsettling and ever-present part of the symphony of his existence as a married father of one. As the narrator puts it, Alex hears a "dog barking hoarsely. The lowing of trucks over bridges. The velvety night in his ears. And the hum, always that hum, which maybe wasn't an echo after all, but the sound of time passing" (340).

Nostalgia of the sort that brings Alex to visit Sasha's old apartment leads Americans like him to fetishize history, to repeat vicious cycles akin to those that Pynchon portrays in *Bleeding Edge*, and to feel owned by the past instead of moving into a more productive future. For instance, Sasha's kleptomania represents the fetishization of history. The objects she piles on a table in her apartment contain her "embarrassments and close shaves and little triumphs and moments of pure exhilaration" (15). They contain the texture of her history and comprise "years of her life compressed" (15). Thus, in collecting the objects, Sasha creates a peculiar archive of her own history—a web that contains pieces of the histories of individuals whom Sasha has encountered and a web that Egan juxtaposes with the sleek, digitally webbed world of the book's final chapter. Yet the objects come to trap Sasha in an identity that she wants to escape, as evidenced by her investment in therapy—even though she presumably continues to steal after marrying Drew Blake and having two children with him, Alison (or Ally) and Lincoln Blake. And Sasha's state of captivity points toward the ways in which all Americans exist as trapped, as owned by their pasts even though those pasts masquerade as liberating or countercultural. The narrator describes the way Alex's history traps him in the book's final chapter: Alex could never "forget that every byte of information he'd posted online (favorite color, vegetable, sexual position) was stored in databases of multinationals who swore they would never, ever use it—that he was *owned*, in other words, having sold himself unthinkingly at the very point in his life when he'd felt most subversive" (316).

Moreover, to quote the narrator's words, nostalgia, for Egan, is "the end" because it leads Americans to romanticize histories that exist as anything but romantic: nostalgia has the capacity to flatten the texture of the past in accord with digital culture's flattening effect, and nostalgia also inhibits American engagement with the present (37). Within the world of Egan's book, romanticized, flattened histories involve, for instance, all of Bennie's band members agreeing that their "gig" at the Mab in San Francisco "went well"—even though it involved cruelty at the hands of the masses and exploitative, privileged individuals (54). According to Egan's portrayal of the event, audience members threw garbage at the stage to signify their petty dissatisfaction with the show, and Jocelyn, one of Bennie's bandmates, performs oral sex on Lou in public while her friend and bandmate Rhea weeps at the horror of the situation (54). Furthermore, as digital-age Americans in particular dwell on and romanticize history, they find themselves in vast states of emotional disconnection that masquerade as connection. For instance, Alex's nostalgic interest in visiting Sasha's old apartment is perhaps indicative of the sense of disconnection he feels from Rebecca, his wife—disconnection that is evident through his inactions as much as his actions. In particular, although Alex obtains a job working for Bennie to publicize Scotty's comeback concert at the former site of the World Trade Center, he opts against telling the news to Rebecca in person. He even fails to T—or text-message—the news to her despite the fact that he imagines the brief and emotionally disconnected message he might write: "*Nu job in the wrks—big $ pos. pls kEp opn mind*" (325).

Americans such as Alex in Egan's book engage in misconceptions of and disconnections from the various presents of the book and how those presents came to be—even in a digital age in which connectivity in the form of quick access to ubiquitous information on the surface seems as though it would work to diminish misunderstandings and misconceptions. For example, Scotty, an anachronism who is trapped in the golden years of his personal history, showcases a misunderstanding of how Bennie developed into a relatively successful New York City–based producer in the music industry. As he wonders, "I want to know what happened between A and B" as points along a linear trajectory of time (101). "*A* is when we were both in the band," he elaborates, "chasing the same girl. *B* is now," a point at which Scotty has lost teeth presumably due to drug addiction, poverty, or both and is working as a janitor to make ends meet (101). Scotty fails to make connections that would allow him to understand how history shapes the present that he sees before him. Likewise, Scotty misunderstands digital technology as a modern phenomenon in a laughable way. He continually makes reference to "X's and O's" in his comparison of human beings to presumably digital "*information processing machines*" and fails to recognize that he is referencing alphabetic symbols for hugs and kisses as opposed to numeric binary code, which he naïvely believes himself to be referencing: ones and zeroes that render digital technology as distinct from analog technology (96). He contemplates whether online and physical, real-world experiences exist as interconnected and essentially the same, but in reality, he fails to grasp either fully. He functions as a sad anachronism in Egan's digital-age fiction—as kindhearted but divorced from reality, and he borders on insanity. He functions as a loser who has evidently lost in his battle with time and also with an unethical capitalist American system that inevitably works to stratify society, to disconnect members of society from one another with the goal of making the elite richer.

Yet Egan underscores that disconnections and misunderstandings emerge even among the most digitally connected of twenty-first-century citizens. Most notably, twenty-first-century New Yorkers' valorization of Scotty in the work's final chapter showcases the ways in which digital Americans exist as disconnected from reality, history, and ethical principles. Indeed, Scotty is a has-been who only gets turnout for his concert because of Alex's "fifty parrots," which create so-called "'authentic' word of mouth" through online means as Twitter tweets might (315). And New Yorkers who have essentially been tricked into believing in the importance of

Scotty as a performer and attending his concert fail to see the ways in which virtual reality functions to manipulate their thoughts and actions in physical reality. Perhaps because flat screens so deeply define their reality, they also fail to see past Scotty's meticulously constructed image at the concert—his "mouthful of porcelain," his cleaned-up veneer, and the authority that his steel guitar gives him as a "long, strange instrument" that conceals his real identity as an "addled geezer" who thinks it makes sense to give a fresh-caught East River fish as a gift to a New York City business executive (326, 334, 334). They fail to see the ways in which American capitalist society has historically disenfranchised Scotty and enfranchised them or that Scotty is, according to the narrator, the "embodiment" of American "unease" after two generations of post-9/11 "war and surveillance" (335). Perhaps more disturbingly, those in attendance at the concert fail to see the problem with Scotty having a concert at Ground Zero. In essence, the bizarre event crystallizes "the new world" that La Doll thinks has emerged, and concomitant with La Doll's vision, Lulu, her daughter and Bennie's new assistant, helps to realize the event as the "living embodiment of the new 'handset employee,'" a metaphorical, natural-born cyborg (to again cite Andy Clark's term) who emerges as a literal cyborg by the more distant future of Egan's "Black Box" (143, 317). They forget or maybe opt to disconnect themselves from the haunting history that defines the site of the 9/11 attacks on the World Trade Center: the fact that over two thousand people died there on the day when, to quote Paul Auster, "the twenty-first century finally begins," and when, as DeRosa suggests in his analysis of Egan's work, a "nostalgia for the future" sets in among Americans because "the absent towers testify to the loss of an exceptionalist teleology" that once set the terms for American conceptions of time (Auster, "Random Notes" 35; DeRosa 89, 101).

Paradoxically, the dramatic disconnection between twentieth-century and twenty-first-century identity and life as Scotty and Lulu exemplify it results in purities of different kinds and thus in a peculiar and problematic connection that Egan probes. As the book's narrator explains, Scotty's concert consists of analog music in the digital age. He sings "ballads of paranoia and disconnection ripped from the chest of a man you knew just by looking had never had a page or a profile or a handle or a handset" (*A Visit* 336). And in singing them, he shapes himself as a "myth" whom "everyone wants to own" in a capitalist nation that remains fixated on myth and commodification (336). He is a hero to the members of his audience even though Egan's readers likely question the heroic status he attains. Lulu, too, embodies purity despite her hybrid cyborg nature and her function as Scotty's literary foil. According to the narrator, the digital photographs on what is presumably her social media page show her as "'clean': no piercings, tattoos, or scarifications" (317). Her body is analogous to flattened digital music as Bennie characterizes it and is pure even though (or perhaps because) she lives most of her life in the flat, online world. Further, she writes in a language that she and members of her generation view as pure, as the title of Egan's final chapter—PURE LANGUAGE"— intimates. She Ts Alex to work out the details of Scotty's concert at Ground Zero, asking, "*U hav sum nAms 4 me?*" and stating, "*GrAt. Il gt 2 wrk*" (321). And in doing so, she believes she communicates in language that is "pure—no philosophy, no metaphors, no judgments" (321). Her texts resemble those that Shteyngart, Roupenian, Ferris, and Foer include in novels they write about digitization. And she believes that the textual phrases exist as an endpoint in the evolution of language that digital technology and human reliance on it together make possible. She fails to see what Egan hopes her book's readers see as they likely have difficulty deciphering the texts: that this kind of language may foretell literature's decline.

Egan, who observes in "Rewiring the Real" that there "needs to be a way for books to live in digital culture or I figure that they really could die," positions literature as a potential catalyst for social change as Shteyngart and Roupenian do. And she, too, positions hybridity as a concept that transcends the digital/nondigital binary and as the alternative to purity in its different problematic configurations. By way of the hybrid form of her book, a form that resembles Shteyngart's novel and Roupenian's story, she invites readers to contemplate new possibilities for hybridity, which, according to Bhabha in *The Location of Culture*, involves liminality: being "neither the one thing nor the other" (49). Perhaps most obviously, Egan's rich prose, which includes the unique written and spoken voices of numerous major and minor characters all in dialogue with one another, functions as the hybrid antithesis of Lulu's pragmatic, so-called pure prose, which, when it appears in Lulu's Ts for work, stripped of fully spelled-out words, contributes to the hybridity of Egan's book. Egan's book, too, exists as hybrid because of its liminal shape between a novel and a short-story collection. As Egan explains in "Rewiring the Real," she never saw *A Visit from the Goon Squad* as a novel necessarily—and she "wouldn't let them write novel on the cover" of the original hardcover because, for her, the book is "not easy to categorize." However, she notes that she learned that "people don't actually buy a book if you don't tell them what it is," so she agreed to market subsequent editions of the book as novels even though readers may find themselves "disappointed" that the book subverts existing genre expectations ("Rewiring the Real").

Moreover, Egan's book exists as hybrid because it combines conventional characteristics of literary texts with characteristics that music aficionados may encounter when they listen to pre-digital-age cassette tapes or vinyl records. The book is divided into two parts—A and B—as tapes and records are divided into sides A and B. Chapters 1 through 6 appear in section A of her book, and chapters 7 through 13 appear in section B. The individual chapters serve as tracks of a recording. And Egan bridges the two sections by titling chapter seven "A to B" to further draw attention to the analog music metaphor she employs and to connect her music metaphor to her theme of time. In this chapter, she describes Bosco's planned album, *A to B*, which treats A and B as points along a linear trajectory of time as opposed to treating them as sides of an album. And this chapter also sheds light on the transition Bennie experiences in his life from being a "cholo" child of an electrician from Daly City, California, who played in the Flaming Dildos in 1970s San Francisco prior to emerging as a married music executive living in a world of cocktail-party-attending, rich, white, and blonde Republicans such as Kathy, his apparent mistress and presumably part of the reason why his first marriage ends (*A Visit* 42). The chapter thus intimates that Bennie's punk-rock roots exist as the lesser-known B-side of Bennie's life even though they precede the A-side in chronological time. His punk-rock roots are akin to the side of a record or tape on which the lesser-known tracks appear—the side that helps define music connoisseurs as such because they tend to appreciate lesser-known B-side tracks most.

Egan, too, hybridizes her book by giving it noteworthy textures that speak to the unique, ribbed texture of a vinyl record, especially as a record's texture contrasts with that of a sleek compact disc. As Morgan explains, "If you zoom in on a single groove of a vinyl record and look at it from the side, the shape would resemble one of the phonautograph drawings" that Édouard-Léon Scott de Martinville developed through the phonautograph he invented in the mid-nineteenth century—the first device on which audio recordings were made. As Morgan continues, playing music on a vinyl record player essentially involves reversing the phonautograph's process: a "thin point, such as a needle, rides along the groove" of a vinyl record to read the texture it contains, and the needle moves "up and down with the peaks and valleys encoded in the record" to produce a sound that corresponds to the textured code. Similarly, Egan's book has textures that can be read like record grooves. As she suggests in "Rewiring the Real," "the texture of the chapters, or the stories as I thought they were, when I wrote them was extremely different and I loved that." She elaborates, "I had three rules that I used to guide me, which were that each chapter should be about a different person, each chapter had to deal with a different texture and feel, and each chapter needed to stand completely on its own" ("Rewiring the Real"). The result is that chapters which represent different eras and people who lived in those eras create dramatically different experiences for readers. For instance, with regard to the texture that time in particular creates, chapter 3, "Ask Me if I Care," captures the feel of the 1970s. And it differs dramatically from chapter 13, "Pure Language," which captures the uneasy feel of post-9/11 America in the digital age and what twenty-first-century Americans living at the time at which Egan publishes her book anticipate that the twenty-first century will become.

Furthermore, hybrid forms that appear within Egan's book echo and underscore it as hybrid in a metafictional way. For instance, in its ninth chapter, it contains an innovative celebrity profile, which Jules writes about Kitty from jail. It meshes together classic features of a celebrity profile that a tabloid magazine might run, black humor, features of personal writing such as those that a journal might contain, and features of academic writing in the form of footnotes. Presumably, Jules writes this profile in a hybrid style in order to thwart the boring nature of the interview he recalls having with Kitty much as Egan writes her book in a hybrid style, ostensibly to counter the potentially deadening effects of digital media on twenty-first-century American society. In other words, Jules works to counter the fact that "Kitty's a snooze" and the fact that the interview lacks an "event" by showcasing his own creativity as a writer—even if he lacks morality and attempts to rape Kitty by the chapter's end, perchance in order to insert an exciting event into the interview and the narrative (*A Visit* 174, 176). Egan's book, too, contains PowerPoint slides that together comprise, in the book's twelfth chapter, an excerpt from Sasha's daughter Ally's "slide journal" (253). As Martin Moling suggests, "By composing the chapter in PowerPoint slides, Egan not only explores new means of creative expression, but also seeks to locate artistic potential in the very 'blankness' of today's digital world" (65). And, to build on Moling's argument, she seeks to explore ways in which impersonal digital-age modes and media that are typically used for corporate purposes can serve personal, noncorporate objectives. She, too, seeks to invite readers to contemplate the problems and possibilities of remediation that results in hybridity, as evidenced by the fact that readers encounter the PowerPoint journal as a remediated piece of writing when they encounter it in print as opposed to its original digital form. They encounter it in a liminal state between a corporate digital document, a personal journal, and a novel chapter or short story. Hence they encounter a composition within her book that exists as a microcosm of or metanarrative within her book because of its hybridity.

The interest in digital writing that Egan showcases by virtue of the PowerPoint chapter speaks to the digital-age-inspired structure of her hybrid book, which takes on features of websites or social media platforms perhaps because she finds these features interesting in her own personal life. As she explains in "Rewiring the Real," she is "fascinated by the fetishization of connection itself." And in the online world, Twitter in particular fascinates her, as evidenced by her serialized Twitter publication of "Black Box," a prime example of Twitter fiction as a genre ("Rewiring the Real"). Indeed, readers of *A Visit from the Goon Squad* come to feel confounded by space and time as they might when they scroll through their Twitter feeds or surf the web. They come to feel like readers of DeLillo's *Cosmopolis* do: perplexed by space and time because DeLillo attempts, in his fiction, to reconfigure Albert Einstein's special theory of relativity. They perhaps feel like they do when they are researching some subject online and then follow a link that takes them deeper into the web and deeper into the time-consuming process of surfing it—the kind of process that Maxine speaks to in Pynchon's *Bleeding Edge* when she reemerges into physical reality from the deep web and asks about the time (77). Whereas the digital revolution both makes more time and diminishes time for digital technology's users, enabling them to attain information and complete tasks with the push of a button but also making them waste time or feel the terrifyingly fleeting pace of modern life, Egan confounds time by making it move both forward and backward over the course of her book to mimic what she calls "that hypertext model" of "following curiosity in a lateral way through a kind of tangle of connections" ("Rewiring the Real"). She tells stories that span roughly a fifty-year period and invites readers to negotiate with her nonlinear tangle of stories.[4]

Readers' responses to confounded time as Egan represents it likely vary and exist as a hybrid complement to the hybrid fiction that she creates. Certainly, Egan's readers may consider connections that exist among characters, most notably Bennie and Sasha, who function as nexus points in the text because they relate in subtle or profound ways with so many other characters. Readers who opt to explore connections such as these as they explore conspiratorial connections in *Bleeding Edge* perhaps see that Egan attempts to invite them to reflect on the social effects of the digital revolution on humanity—on the illusion of connection that digitization creates through social media platforms such as Facebook, which Egan critiques much in the way that Zadie Smith does in "Generation Why?" Egan views Facebook as the aesthetic opposite of her dissimilarly textured tales: as "dull" and consisting of profile versions of life that have "horrific aesthetic conditions" like "huge Soviet" apartments in which everyone's "cell looks exactly the same" ("Rewiring the Real"). Throughout her book, Egan stresses the existence of these connections among characters, be they connections in the physical or burgeoning digital world. As the electrical engineering PhD student Bix prognosticates based on his knowledge of the internet in chapter 10, "Out of Body," "The days of losing touch are almost gone," even if digital connections never equal or resemble in nature authentic human connections (*A Visit* 203). And the narrator of the "Safari" chapter foretells a similar future of American and global disconnected connectivity in an oracular voice that peppers Egan's book. As the narrator puts it, the members of the safari will remember their experience and feel compelled to "search for each other on

Google and Facebook, unable to resist the wish-fulfillment fantasy these portals offer," but they will ultimately realize that "having been on safari thirty-five years before doesn't qualify as having much in common" (71).

Alternately, some readers such as the literary critic Heather Duerre Humann may feel compelled to "retroactively" piece "together details to form a coherent narrative" (91). They may want to spend time in the weeds of the past in Egan's fiction to discover which chapter comes when and what characters' stories would look like if Egan had strung them together chronologically. They may want to create connections among the various pasts, presents, and futures that Egan presents because the process of making connections exists as a valuable process of meaning making, a key concern for Pynchon as well. As Humann discovers in her own critical untangling of Egan's tangle, Sasha feels lost amid the objects she steals because she fails to understand her own history: she fails to understand that she steals because of "traumas she's endured over the years" (Humann 90). She neglects to see that watching her father, Andy Grady, physically abuse her mother, Beth, at Lake Michigan when she is a child perhaps leads her, first, to run away to become a "hooker and a thief" in Naples and, second, to become a kleptomaniac when she returns to the United States (Egan, *A Visit* 204). The detangling efforts in which Humann engages may even lead readers to connect fiction with reality: to explore ways in which the United States moved from A to B, to employ Egan's metaphor. Readers may desire to understand the way in which the hybrid world of punks and business executives that she chronicles develops into the sleek yet anxious, digital, post-9/11 world in which Americans of Egan's real-life present live.

Alternatively still, readers of Egan's hybrid book may opt to celebrate it as a socially conscious work of art that underscores the potential that hybridity as disorder holds because disorder presents an opportunity for a new order to arise: it offers potential for Americans to realize a better hybrid future than the corporate and dystopian post-9/11 future that her fiction portrays. Readers may look at Egan's book as a work of art just as Ted, Sasha's art historian uncle, looks at the Orpheus and Eurydice relief while in Naples with the charge of finding the lost Sasha for his sister. But instead of failing to make the connection between art and life as Ted fails to make it, Egan's readers may look to her book as an aesthetic object that intends to give shape to human ethics and actions. Indeed, the image with which Egan ends her book—that of a woman who is "young and new to the city" and who is "fiddling with her keys"—might be read as symbolic (340). On one hand, it can signify the cyclical nature of history. The woman appears as the new Sasha at a moment when Alex and readers alike have come full circle by returning to her apartment, which Egan portrays in the first chapter of her book. Readers come back to the beginning as they do in Pynchon's *Bleeding Edge*. On the other hand, however, this moment, too, is symbolic of a new beginning. The moment is filled with potential because this new woman is *not* Sasha and has her whole life ahead of her. And Sasha's history should not necessarily influence readings of this woman, whose story remains, in Egan's book, poignantly unwritten. Instead, Egan leaves the woman to write her story by virtue of the life she opts to live. Similarly, to reference and build on Moling's argument, she complements her effort to expand on "the scope of the art of the novel" with the goal of providing it "with a future" and with a concern for the future that will manifest itself in historical time (53). She leaves Americans who read her book to write a counternarrative in emergent history. She leaves them to write a counternarrative to the dystopian future that she presents—a narrative that is not rooted in problematic incarnations of nostalgia such as those that her and Pynchon's characters experience.

Ideally, Egan's readers will write a hybrid narrative by way of living their lives in historical time, and this narrative will resemble Egan's book, functioning as a B-side (in the musical sense) to twentieth-century exceptionalist American history that came to an end on 9/11. Ideally, too, twenty-first-century American history will be comprised of a metaphorical set of more humble tracks: acts that, for better or worse, lack the glitz of those that defined the American Century, but acts that true connoisseurs of history will recognize as great just as music connoisseurs recognize B-side tracks on albums as impressive. This emergent twenty-first-century historical narrative will certainly involve digital technology because there exists no way to run history in reverse—nor is Egan a Luddite who desires the erasure of technological progress. Hence this narrative will almost certainly involve metaphorical or perhaps actual cyborgs (such as Lulu in "Black Box"), depending on the directions in which technological developments move. However, according to the ideal toward which Egan gestures, this narrative will also involve the development of and a reliance on a punk, countercorporate ethical compass that gives texture to the fabric of American life and that speaks to Pynchon's vision of the twenty-first-century badass in *Bleeding Edge*. As a result, even when the "answers" to ethical questions appear to be "maddeningly absent," to reference the narrator's articulation of Alex's reflection about his past experience with Sasha, Americans will have a reliable means by which to probe history to find answers and make connections in order to forge a more ethical future (311). And they will, with any luck, achieve what Sasha cannot by the end of Egan's fiction: a means by which to find redemption. They have, through making a connection between socially conscious art and life and altering their actions according to reflections that Egan prompts, the opportunity to redeem the United States from its past failures. They thus have the opportunity to transform the hum of time passing amid stasis into a more melodic tune through the societal progress they help enact.

CHAPTER 3

The Digital Divine in Joshua Ferris's *To Rise Again at a Decent Hour* and Jonathan Safran Foer's *Here I Am*

American history as Thomas Pynchon and Jennifer Egan consider it is inevitably entwined with religion, which Frank J. Lechner and John Boli see as the peculiar source of digitally mediated globalization. As Lechner and Boli explain, "Long before the current phase of globalization" that digital media make possible, "religious communities globalized" (387). Lechner and Boli elaborate, explaining that each "religion had its own kind of mobile messenger, each its own universal message, and each its own impulse to include new adherents" (387). Yet the contemporary connection between digitization and religion as political phenomena is complicated, as in part demonstrated by Neil Postman's observations in *Technopoly: The Surrender of Culture to Technology*. Postman traces the beginnings of American Technopoly, or totalitarian technocracy, to the Scopes Monkey Trial, which pitted religion against science and technology and which functions as "an expression of the ultimate repudiation of an older world-view" that Christian fundamentalism in the United States embodies (50). By contrast, Thomas L. Friedman gestures toward symbiosis between digitization and religion. As Friedman suggests by virtue of the title of his 2003 *New York Times* op-ed "Is Google God?", which anticipates and shares references with *The World Is Flat*, digital media may well be supplanting the functional role of divinity. In his article, as in his 2005 book, Friedman quotes a vice president of the then-new wireless provider known as Airespace, Alan Cohen, who observes that "Google, combined with Wi-Fi, is a little bit like God." As Cohen continues in his interview with Friedman, "God is wireless, God is everywhere and God sees and knows everything. Throughout history, people connected to God without wires. Now, for many questions in the world, you ask Google, and increasingly, you can do it without wires, too" ("Is Google God?"). And this sense of Google as God speaks to David F. Noble's *The Religion of Technology: The Divinity of Man and the Spirit of Invention*, which argues that "the present enchantment with things technological" is "rooted in religious myths and ancient imaginings" (3). "Today's technologists," Noble continues, are driven by "distant dreams, spiritual yearnings for supernatural redemption." Hence in the new millennium, to borrow Noble's words, humanity bears "witness to two seemingly incompatible enthusiasms, on the one hand a widespread infatuation with technological advance and a confidence in the ultimate triumph of reason, on the other a resurgence of fundamentalist faith akin to a religious revival" (3).

This chapter examines these competing and at times cooperating enthusiasms in Joshua Ferris's *To Rise Again at a Decent Hour* and Jonathan Safran Foer's *Here I Am*, novels that turn to aspects of organized religion and specific religious texts as a means by which to understand or address problems of digital-age, hybridized life, to build on Donna Haraway's and Homi K. Bhabha's respective considerations of hybridity. In the first of the two parts of this chapter, I consider Ferris, who sees digital devices and social media as unnecessary in his professional life even though his writing expresses a clear interest in them. As Ferris acknowledges in a 2014 interview with Darren Richman, "I have a strange relationship with technology," and "I live a pretty analogue life" ("Joshua Ferris: The Writer on Hard Work"). He elaborates in a 2015 interview with Mary Laura Philpott: "I'm not on Twitter and I don't have a Facebook page, but my attitude is less about scorn and more about the need to write books. If I were tweeting all the time I'd probably get obsessed with Twitter, so it's really just a matter of professional necessity" ("Joshua Ferris: Why Comedy"). Moreover, as he tells Edie Greaves in a 2014 interview, "I don't like writing on a computer" ("The World of Writer Joshua Ferris"). Yet technological devices appear with ubiquity in the modern American workplace of *Then We Came to the End*, Ferris's debut novel, which loosely addresses his work experiences in a Chicago ad agency. He sees these devices as shaping companionship among colleagues and quelling the deadening effects of corporate interests.[1] And he sees ways in which digitization allows for transcendence of the mundane aspects of everyday contemporary life. As he tells Christopher Bollen in a 2014 interview, which echoes the words of Cohen in Friedman's "Is Google God?", the internet is "a kind of deity" for the ways in which it appeases loneliness ("Joshua Ferris"). "If you spend enough time on it," Ferris continues, "there's not really that nagging sensation of existential despair; it erases it very effectively. And it's monolithic. And thanks to it, we've got the ways to linger there after death—your blog exists afterwards, your e-mail exists afterwards." Indeed, the internet has the potential to provide a religious experience for secular individuals such as Ferris, who says to Jonathan Lee in a 2014 interview that he "wanted a religious community" for himself, perhaps because, growing up, he "didn't have one" ("Always on Display").

My analysis of *To Rise Again at a Decent Hour*, Ferris's most sustained meditation on digitization, considers Ferris's exploratory literary philosophy of technology that builds on ideas he presents in *Then We Came to the End*, which Alison Russell sees as showcasing Ferris's interest in "Emersonian (and very American)" ideals about individualist identity and an apparently conflicting interest in community, or "connectedness in an increasingly disconnected society" (319). In telling the story of the paradoxical Paul O'Rourke, a Manhattan dentist who obsesses over religion and the digital world while proclaiming atheism and neo-Luddism, Ferris portrays ways in which religious impulses persist as part of American identity, particularly for Americans who deify digital devices and media in ways that are evocative of Friedman's romanticized representation of digitization in *The World Is Flat*. Over the course of the novel, Ferris maps out a tortuous matrix of connections between religion and digitization as well as paradoxes of digitization that underpin contemporary American life. He spotlights ways in which digital devices and media provide opportunities for the transcendence of everyday physical experience while disconnecting individuals from communities. He draws attention to ways in which digital devices and media flatten and fracture individual identity, a phenomenon that he illustrates through his portrayal of Paul's experience with having his online identity hijacked by Grant Arthur, the founder of Ulmism, a fictionalized religious faith that is rooted in devout doubt in God's existence. And he reveals ways in which socially constructed and perhaps unreliable digital texts, which are analogous to traditional religious texts, both facilitate the acquisition of information while inhibiting the development of knowledge and altering notions of truth and reality. In turn, he suggests that although conventional novelists may lack political power to radically transform the flat world's social order, they can still prompt readers to consider the persistent problems that digitization presents. And he intimates that in order to find meaning amid digital malaise in twenty-first-century life, twenty-first-century citizens must accept digitization and the notion that virtual, spiritual, and material reality intermingle in inextricable ways. He intimates that these citizens, like authors, must work to develop multiliteracies to emerge as more incisive critical thinkers and readers of matrices of truth and fiction in digital, physical, and metaphorical texts. And they must find ways to disengage with potentially toxic digital and religious concerns in order to engage in meaningful work for the social good. Through meaningful work, they can challenge digital-age disconnection by building community locally and across national, physical, and virtual borders. And they can face and perhaps embrace the notion that, as in Ferris's novel, the uncertainties of the contemporary moment may well inform the development of a more genuinely interconnected future in the globalized world.

Whereas Ferris does not identify with a specific faith, Foer as a Jewish American author has consistently addressed Judaism as well as digital-age multimodality in tacit and overt ways in his often experimental literary works. In the second part of this chapter, I examine how Foer sees contemporary American and perhaps also global twenty-first-century life experiences as inherently involving negotiations with ideas that are central to Jewish history. For Foer, both religion and digitization function as forces that hybridize individuals' identities because they have the capacity to fragment life experiences and notions of what counts as the self. For instance, in *Everything Is Illuminated*, Foer represents the fragmented experience of an American Jewish vegetarian who stands in stark contrast to his eastern European Jewish ancestors and to the largely anti-Semitic 1990s Ukrainians he encounters as he attempts to investigate his family's Holocaust-era history. By producing a fragmented text that interweaves with Foer's own personal biography,[2] Foer invites readers to piece together a narrative much as his novel's protagonist must develop a narrative of selfhood through the fragments of history he encounters, most notably the photograph of the woman who purportedly saved his grandfather from the Nazis. Along the same lines, Foer showcases the hybridizing force of digitization. Digital media communication functions as a clear influence on the formal features of "A Primer for the Punctuation of Heart Disease," a multimodal short story that introduces atypical punctuation marks which challenge conventional notions of what counts as a print text by denoting in imagistic form the unspoken rhetoric that develops among family members with a history of heart attacks. Digital media also function as a clear influence in *Extremely Loud and Incredibly Close*, a multimodal novel that ends with a series of images that the protagonist finds on the internet as he works to come to terms with the loss of his father in the attacks of 9/11, a modern-day atrocity that displays political tensions between the capitalist West and Islamist fundamentalism. And, perhaps most notably, digitization clearly influenced the creation of *Tree of Codes*, an X-Acto-knife-edited, book-art version of Bruno Schulz's *The Street of Crocodiles* that features a different die-cut on each page. As Kiene Brillenburg Wurth explains in her analysis of the book, Foer "invokes concepts central to computer programming" through his title, and he makes readers "sensitive to what it means to read a book, a page—not just a text—in an age where Kindle and other tablets are undoing the crucial dimension of page design" through the book's content.

My analysis in the second section of this chapter complements Wurth's reading of *Tree of Codes* as an artistic response to digitization. I consider *Here I Am* as Foer's most philosophical meditation in fiction on the practical effects and ethical implications of digital engagement as well as "Foer's most Jewish novel," to cite Karen Heller's interview with Foer ("Jonathan Safran Foer Interview"). I argue that by depicting domestic scenarios involving the American Jewish household of Jacob Bloch, a writer, and his wife, Julia, an architect, and also in representing natural and politically cataclysmic events involving Israel, Foer dramatizes the fragmentation of life experience in an increasingly digital world that Friedman idealizes as interconnected in *The World Is Flat*. By way of a Minecraft-esque digital platform known as Other Life, which Jacob and Julia's eldest son, Sam, engages with as he avoids the fact of his great-grandfather's suicide, his parents' looming divorce, and fallout from an accusation that he uttered a profanity, Foer establishes a metaphor for what he views as the central problem of the digital age. He invites his readers to see ways in which modern-day Americans live an assortment of other lives and thus exist as paradoxically present without presence in their real lives. He likewise invites them to see ways in which the other lives that literary and religious texts afford function as antecedents to the other lives that the digital world offers. Ultimately, with rhetorical purpose, Foer highlights the distinction between fragmented digital-age life and the kind of life that the biblical Abraham models through showing a paradoxical presence for both God and Isaac, his son and prospective sacrifice in Genesis 22, the source of the title *Here I Am* and a text that works with Genesis 32 to illuminate the philosophical center of Foer's novel. As Foer sees it, although twenty-first-century citizens might not always successfully erase rifts that emerge in their lives as a result of the ubiquity of digital devices and media, they might, with help from these broadly relevant and applicable Jewish narratives, wrestle with the prospect of maturity in accord with the meaning of Israel in a way that is evocative of Ferris's portrait of the utility of work. They might, too, successfully emerge as more integrated twenty-first-century citizens who can join together in a multifarious chorus that has the capacity to function as Foer's novel does. It can reveal philosophical complexity in what it means to live in ethical ways in the ever-digitizing present and future.

Paradoxes of Religion and Digitization: Working to Find Meaning in Joshua Ferris's *To Rise Again at a Decent Hour*

Perhaps as a result of the colonial notion that America exists as a city upon a hill, to reference John Winthrop's seventeenth-century appropriation of Jesus's words from the Sermon on the Mount,[3] religious impulses persist as part of American national identity despite the ways in which scientific and technological developments have come to pervade everyday American life. In *To Rise Again at a Decent Hour* (2014), the focus of the first part of this chapter, Joshua Ferris dramatizes the problems and possibilities of contemporary manifestations of these quintessentially American religious impulses through his portrayal of Paul O'Rourke's various devotions and desire for faith. For instance, Paul experiences a religious impulse that leads him to a fanatical devotion to Red Sox games, which he watched as a child with his father, Conrad, prior to his father's suicide. And through a string of failed relationships with women of different religious heritages, Paul experiences a religious impulse to explore possibilities for developing a more traditional religious identity through conversion like that which his biblical namesake experiences in Acts of the Apostles 9. Paul toys with the notion of converting to Catholicism as a result of his relationship with Samantha Santacroce because he sees her Catholic faith as a means by which he can escape his own trauma and family history. As Paul explains, "The Santacroces were a picture-perfect family of Catholics whose tidy garage, sturdy oak trees, and family portraits through the ages would absolve all the sins and correct all the shortcomings of my childhood" (*To Rise Again* 53). Through conversion, he hopes to "condemn abortion and drink martinis and glory the dollar and assist the poor and crawl upon the face of the earth with righteousness and do everything that made the Santacroces so self-evidently not the O'Rourkes" (54). Similarly, Paul entertains the notion of converting to Judaism as a result of his relationship with Connie Plotz, his ex-girlfriend. As a nonpracticing Jew who comes from a conservative Jewish family, Connie affords Paul another chance to escape the fruitless confines of his identity during the duration of their romantic relationship. As Paul explains, "I wanted to be a Plotz. I wanted to be a Jewish Plotz who sat shiva and went to shul and made babies with Connie behind the bulwark of safety that was the Plotz extended family" (even though Paul eventually decides that he wants no children because he fears they will die) (107).

Ferris, too, suggests that religious impulses lead Americans to use the internet because it sustains a similar social function to religion in contemporary American life. Much as religion connects individual believers through institutions such as churches, synagogues, and mosques, the internet connects digital citizens virtually, complementing or perhaps even supplanting their physical communities, for better or worse, a point that Dave Eggers makes in *The Circle*. In *To Rise Again at a Decent Hour*, Ferris complements descriptions of Paul's efforts to convert to Catholicism and Judaism, respectively, through his significant others with descriptions of Paul's conversion into a user of digital and social media. Initially a skeptic of digitization or digital atheist, Paul resembles Jennifer Egan's Scotty Hausmann in *A Visit from the Goon Squad* in that he attempts to "opt out" of digital life by avoiding social media platforms such as Facebook (Ferris, *To Rise Again* 125). He remains a devotee of VHS tapes and stockpiles VCRs in an effort to fend off the digital future. And he sees himself as "feeling more disconnected than ever" because of digitization and wonders from where "this idea of greater connection" in the digital age comes (32). In his quasi-atheistic resistance to modernity, he paradoxically even resembles fundamentalist believers who, according to Karen Armstrong's *The Battle for God: A History of Fundamentalism*, "are conducting a war against secular modernity" (vii). Yet Paul converts to internet addiction after his identity is hijacked by Seir Design, a digital front for Grant Arthur, the founder of the fictionalized Ulmist faith, about which Paul initially learns from an eccentric patient in his dental practice. In creating a website for Paul's office and a Facebook page for Paul, both of which quote from Ulmist texts that provide anti-Semitic counternarratives to the Bible, Grant embeds Paul in a religious community online. He sends Paul down a metaphorical rabbit hole that corresponds to the conspiratorial rabbit holes of Pynchon's *Bleeding Edge*. Specifically, Paul engages in identity exploration and internet searches involving Ulmism, a faith that prescribes doubt in God as a central tenet and is thus evocative of what Alister E. McGrath and Joanna Collicutt McGrath term atheist fundamentalism.[4] As Paul explains, "I tried my best to fend off the Internet's insidious seduction, until at last all I did—at chairside, on the F train, supine upon the slopes of Central Park—was gaze into my me-machine and lose myself on the Internet" (*To Rise Again* 74).

Along the same lines, Ferris suggests that digital devices and social media serve a godlike function in that they offer individuals and communities apparent opportunities for transcendence, a notion that Noble anticipates in *The Religion of Technology*. In Ferris's novel, opportunities to transcend the physical world through digital means resemble those that religious believers experience or imagine they will experience through faith. Most notably, digital devices and social media allow Americans to transcend through virtual reality the mundane nature of everyday work in offices, which Ferris describes in *Then We Came to the End* and in "More Abandon (Or Whatever Happened to Joe Pope?)." Through online engagement with websites that can "live on" for "all eternity" through caching, Paul transcends boring tasks that he once viewed as revelatory, as illustrated by his remark that the mouth of a girl with whom he had fallen in love in his youth functioned as "a revelation" (*To Rise Again* 77, 77, 195). He transcends the everyday monotony of his job, a word evocative of the biblical Book of Job, to which Ferris draws attention in the novel's epigraph: "Ha, ha," a quote that spotlights both the absurdity of the trials that the devout Job endures at God's hands and the absurdity of Paul's repetitive work of reminding patient after patient to "please floss, flossing makes all the difference" (*The Bible: Authorized King James Version*, Job 39:25; Ferris, *To Rise Again* 4). Indeed, throughout *To Rise Again at a Decent Hour*, Paul emails Seir Design incessantly while at work instead of working. He notes in an atypically self-aware message, "I'm on the Internet all day long and I'm not even in IT" (73). For Paul, finishing sutures prompts him to "feel the exile of age in America" (122). That feeling juxtaposes with the transcendence he experiences when he uses his me-machine, which is godlike in the ways it answers questions that occupy Paul's imagination and supplants God as an inevitably mute and thus Barthian conversation partner in prayer.[5]

Yet Ferris reasons that digital devices and digital and social media paradoxically also disconnect humans from one another, rendering the transcendence they provide as problematic. They disconnect contemporary individuals much as religion may disconnect believers of different kinds from one another in a post-9/11 American nation that is shaped by anti-Muslim sentiments and xenophobia. They inhibit communication and connection even though, in Paul's words, no "invention in the world, not the printing press or the telegraph, not the post office or the telephone, had done more to get people communicating than the Internet" (124). Specifically, as a result of his digital engagements, Paul emerges, in Mohsin Hamid's words, as "present without presence" at work (*Exit West* 40). At best, he goes through the motions of his job like the disingenuous religious believers in Foer's *Here I Am* go through the motions of faith. At worst, however, Paul's work comes to function as a virtual distraction from his online escapades, resulting, for instance, in his request for a stool sample from one of his patients. Later in the novel, Paul even notes that he has "no fucking clue" what he is "supposed to be doing" for a patient because the digital world distracts and disconnects him so deeply from his profession (*To Rise Again* 213). Along the same lines, and illustrating the ways in which television functions as a cultural antecedent to smartphones in shaping contemporary notions of what counts as digital screen culture, Paul's mother complements churchgoing with disconnection from reality through television following her husband's traumatizing suicide. In Paul's words, "in what I imagine now to be one of many desperate attempts to organize her response to the inconceivable," she "cycled through a series of churches" and then found her way "back home again to sit on the sofa and mourn in the everyday way of most Americans: in the communal privacy of the TV" (117). Later in life, albeit for a different confluence of reasons, Paul's mother appears as entirely disconnected from reality when Paul visits her in the Sarah Harvest Dodd Home for the Elderly in Poughkeepsie, New York. As Paul describes her, she lacks a "functioning brain" as well as a functioning body due to her old age and perhaps, too, due to the lasting effects of her past emotional trauma (263). Instead of interacting with her son, she merely "hummed a lot and stared at the TV" (263). She appears as a symbol of humanity's vegetative state as a result of overengagement with screens.

Furthermore, Ferris intimates that while fostering quasitranscendent degrees of connectivity, digital devices and social media also fracture identity and flatten humanity, countering the effect that transcendence produces. Notably, while masquerading as Seir Design online and thereby bifurcating his own identity through digital means, Grant draws attention to problems involving identity in the globalized, interconnected world. By fictionalizing a false and alienating online biography for Paul and by asking Paul how well he knows himself after Paul questions the content of the online biography, Grant prompts Paul and Ferris's readers to contemplate ways in which the digital world allows for the creation of selves that parallel or perhaps even run counter to real-life, physical selves (78). He prompts Paul and his readers to see that the digital self is a carefully curated and rhetorical representation of the real self, a point that Kristen Roupenian aestheticizes in "Cat Person" and that Sherry Turkle explicitly makes when she notes that through digital means, we "hide from each other even as we're constantly connected to each other" and "are tempted to present ourselves as we would like to be" (*Reclaiming Conversation* 3, 4). Grant likewise prompts Paul and Ferris's readers to contemplate ways in which the digital world has the capacity to circumscribe human identity and flatten the texture of humanity through purposeful rhetorical curations and other incidental means. As Paul suggests in a technophobic rant early in the novel, "I was a dentist, not a website. I was a muddle, not a brand. I was a man, not a profile" (*To Rise Again* 32). Referencing an apparently omnipotent digital force that has the power to flatten his human life, he notes that they "wanted to contain my life with a summary of its purchases and preferences, prescription medications, and predictable behaviors. That was not a man. That was an animal in a cage" (32). And as Paul explains later in the novel in an angry email to Seir Design that illustrates through an autocorrection the absurdity of online communication, "I will not be contained by my news feeds and online purchases, by your complicated algorithms for simplifying a man. Watch me break out of the hole you put me in. I am a man, not an animal in a cafe" (133).

The ways in which the internet allows for false representations of individuals' identities speaks to a paradox involving truth and knowledge of verifiable truth that emerges out of a historical consideration of the information revolution, which begins with the invention of the printing press and involves the interplay of technology and religion from the outset. As Thomas J. Misa points out in *Leonardo to the Internet: Technology and Culture from the Renaissance to the Present*, the "invention of moveable type for printing" as Gutenberg's printing press employed it prompted "an information explosion that profoundly altered scholarship, religious practices, and the character of technology" (19). And in the mid-1450s, it famously allowed for the dissemination of knowable truths by means of the printing and dissemination of printed texts, for instance the Gutenberg Bible, a Latin version of the Hebrew Old Testament and the Greek New Testament, which exemplified the influence of new technologies of old. Moveable type likewise allowed for the printing and subsequent dissemination of Martin Luther's theses, which "helped usher in the Protestant Reformation" (Misa 22). It thereby allowed for a textual proliferation that Ferris values as a reader and author, and it also allowed for the propagation of ideas and ideologies that texts contain. In other words, it stimulated intellectual development. Moreover, it set Western civilization on its course toward modern-day problems of epistemology that concern Gary Shteyngart in *Super Sad True Love Story* (as evidenced by his title's attention to truth). Ferris references such problems via his discussion of Paul's visits to Carlton B. Sookhart's Rare Books and Antiquities shop, where Paul asks Sookhart, the business's proprietor, whether he has "heard of something called The Cantaveticles," the collection of so-called cantonments to which Grant makes reference when he poses as Paul online (Ferris, *To Rise Again* 108). Through his description of the cunning Sookhart, a likely forger who contrasts the online account that Grant attributes to Paul with the biblical account, Ferris deconstructs the binary involving authenticity and print on one hand and fabrication and the internet on the other. He suggests that digitization does not impede truth or authenticity and that print culture does not inherently produce them. Hence he indicates that digitization does not make life worse per se; it only sheds greater light on problems of truth and authenticity that already exist. As the narrator explains, "Sookhart had brokered many high-profile transactions over the years"—even one involving a Gutenberg Bible—but in "the late nineties, his reputation suffered a blow when a private collector and thermal chemist accused Sookhart of forgery," which Sookhart likely commits again near the end of Ferris's novel when he offers Paul a Yiddish copy of the Cantaveticles (108–109, 109).

Ferris alludes to the noteworthy connection between the paradox of knowing that emerges as a result of the invention of the printing press and the paradox that emerges as a result of the commercialization of the internet, which accelerated access to and thereby changed the nature of information when it transcended its militaristic, Cold War American roots in the 1990s. In part, Ian Sample's 2019 *Guardian* article acknowledges this paradox of knowing and also implicitly challenges Friedman's notion that the flat world levels the playing field to create equal access and social equality. As Sample observes, the internet leads contemporary citizens to view the world as *literally* flat, as evidenced by new research that identifies You-

Tube as "the prime driver for a startling rise in the number of people who think the Earth is flat." And Michael Patrick Lynch illuminates this paradox of knowing in different terms, suggesting, in *The Internet of Us*, that information technology is "expanding our ability to know in one way" while "actually impeding our ability to know in other, more complex ways" (6). As Ferris suggests, in ways aligning his own literary argument with the argument of Lynch's book, the internet as the new definitive technology of the information age obscures knowledge and truth because it presents such a vast amount of information, a concept toward which Ferris gestures throughout his novel by way of the many (largely failed) searches that different characters conduct. Further, for Ferris the internet functions to reveal truth in unexpected ways that transcend the mere presentation of mass amounts of information. Ferris exemplifies this function through the reference he makes to the revelatory qualities of the Streisand Effect. Named after Barbara Streisand, who famously attempted to censor online information about her lavish Malibu home, the Streisand Effect results when an individual attempts to suppress information online. As Paul describes it in relation to his own desire to remove online material posted by Grant, "once people knew I was trying to suppress something published on the Internet, they would actively seek it out to see what the fuss was all about, which would create a negative feedback loop, more attention drawing yet more attention" (*To Rise Again* 131). In other words, there exists an impulse to make information free through digital means, as illustrated, for instance, by controversies involving Wikileaks (the media organization via which Julian Assange exposed unsettling U.S. government secrets) and debates about net neutrality (the notion that internet service providers should provide equal access to online information to all users and not slow down, block, or charge extra for any information access).

Much as Ferris juxtaposes historical and modern-day problems involving knowledge, he juxtaposes historical and contemporary conceptions of reality as a social construct that is informed by textual production. He thereby showcases that digitization sheds greater light on problems of reality but perhaps does not impede notions of the real more so than print culture does. On one hand, he draws attention to the social construction of reality in general and religious reality in particular by way of a metafictional moment in his novel. Specifically, he asks his readers to contemplate a question that Paul asks Sookhart: "how many people" it takes to "make a thing [. . .] real" (181). In turn, he asks his readers to contemplate Paul's follow-up question for Sookhart: "How many people do you need to say that a system of belief is a bona fide system?" (182). In asking his readers to contemplate these questions, Ferris invites them to ponder the construction of real, historical, print books, particularly religious texts such as the Hebrew and Christian Bibles, and the realities that these books help shape by virtue of their inevitably rhetorical existence. He invites readers to consider the fact that the Bible is not just one text, plain and simple, but a cobbled-together, unstable, and still-changing work. It is a much translated and highly political anthology of many texts written in different periods by different authors, and Ferris exhibits this reality by virtue of making reference to the "vigorous debate" that has existed regarding "Job's authorship" (235). As scholars of religion suggest, these authors held different worldviews and motivations, and, to quote Sookhart, they turned to different "urtexts" and "prototexts" such as *Enuma Elish* or "the *Epic of Gilgamesh*" to develop their own rhetorical narratives (236). Ferris likewise invites his readers to consider the political circumstances that influence what comes to count as canonical as opposed to apocryphal. As Michael Coogan explains in *The Old Testament: A Very Short Introduction*, "Stabilizing the contents of the Bible in what came to be called the canon, an official list of the authoritative books, was a gradual process," and despite "naïve views to the contrary, the Bible was not handed down by God as a complete package but was the result of a series of decisions made over the course of centuries by the leaders of different religious groups" (7, 11). Hence religious faiths that are based on religious texts are rooted in decisions about inclusions and exclusions. They are also rooted in various readings and interpretations of religious texts—approaches to making meaning that inevitably privilege some ideas, individuals, and groups while disenfranchising others.

On the other hand, Ferris contemplates the contemporary social construction of reality in general and religious reality in particular through his representations of online textual production. And he suggests that notions of what exists as real or authoritative in the hybrid, globalized, twenty-first-century world and even of what counts as that world rely on complex processes of human collaboration and consensus much as the historical religious texts to which he alludes rely on them. For instance, Ferris portrays Grant as continually editing the online texts he posts when he is posing as Paul, altering Paul's sense of reality and also altering reality for all his readers with each of his edits. And Ferris also portrays invisible corporate forces at Facebook as paradoxically contributing to what counts as material reality in the unwebbed world that social media complement. He describes the internet giant as representing and thus endorsing Ulm as a "religious affiliation" on the Facebook page that Grant constructs for Paul despite Paul's sense that Ulmism is not an actual faith (*To Rise Again* 125). More to the point, Ferris describes the way Wikipedia crowdsources a representation of reality via the digital entries its editors opt to include. These entries are related to authoritative entries of old that were published in print encyclopedias. Although, according to Paul's limited understanding of how Wikipedia works, "trekkieandtwinkies, one of Wikipedia's self-appointed editors," initially nominates a page on Ulmism for deletion "on the grounds of an insufficient something or other," Ulmism eventually manifests as a real faith, so to speak, in large part because of Wikipedia's editors (162). Paul describes encountering the new reality that Wikipedia establishes: "I was in the Thunderbox when I came across the fourth, or maybe the fifth, iteration of the Wikipedia entry for 'Ulm,'" but unlike "earlier attempts, this one had been approved for publication by Wikipedia's editors" (210).

Through meditations on information, knowledge, reality, and textual production that *To Rise Again at a Decent Hour* facilitates, Ferris effectively illuminates the means by which religion has worked in conjunction with digital devices and media over the course of history to construct the post-truth era, to reference a term that Ralph Keyes coins in his 2004 book.[6] In other words, Ferris illuminates the way in which emotions and beliefs have come to exist as more valuable than verifiable facts, particularly in relation to political debates. His text is evocative of the notion that, to borrow Cotton Mather's term, the religiously charged invisible world,[7] or the world in which unseen demonic and divine forces operate, sets the United States on its course to the post-truth present. His text, too, is evocative of the notion that staunch faith in the invisible world renders inevitably unreal objects of faith into factual, real phenomena by 1910 with the publication of the first of *The Fundamentals* and thus the birth of fundamentalism. In turn, his novel indicates ways in which modern-day Americans on the internet have a proclivity for reading online texts literally or unquestioningly and may accept these texts as unequivocally true—even if they have no personal history with fundamentalism and even though the truth of these online texts should remain questionable. Although Ferris portrays Grant as perhaps dubious in constructing fake biblical passages that strategically echo biblical rhetoric about the covenant and the construction of "a great nation," he focuses his critique more so on Grant's text's readers, who span the religious and secular spectrum to include, for instance, Paul and also Paul's chief hygienist, Betsy Convoy (60). Indeed, as a devout Catholic, Betsy should have a penchant for embracing the mysteries of faith as opposed to staunch literalisms that characterize fundamentalist Protestantism. She should appreciate the unknown and unknowable as characters throughout DeLillo's oeuvre appreciate them. Instead, however, at least in this moment, her remarks suggest that she is at risk of being brought into the problematic fold of digital-age confidence and certainty. She observes that Grant's passage sounds "very stern," and not knowing exactly the book from which it comes, she guesses that the text comes from the "Old Testament"—even though she in actuality knows nothing of its origins (62). She thereby illustrates the slow and steady process via which assumptions morph into misinformation that masquerades as fact in twenty-first-century life. Too, she illustrates that the desire to verify the truth of what gets said in everyday exchanges is as immaterial—in both senses of the word—as the virtual realities that digital media proliferate.

Ferris suggests that regardless of the problems that post-truth American life ushers in, contemporary Americans and citizens of the globalized world must reconcile with the matrix of benefits and detriments of digitization because, like notions of spirituality that religious believers engage with, digitization exists as part and parcel of reality. To reference Eggers's *The Circle*, there exist physical and invisible subspaces where data are stored such as the space to which the tech savant Kalden brings Mae Holland, *The Circle*'s protagonist, to show her Stewart, a massive data storage unit. Likewise, we have to engage with the digital world in real ways because we have "reached a time when the phrase 'digital rhetoric' is redundant," to reference Gurak and Antonijevic's words in "Digital Rhetoric and Public Discourse" (497). As Gurak and Antonijevic astutely elaborate, no rhetoric is "beyond the realm of the digital" (497). And, illustrating the pervasive ways in which the digital pervades very real rhetorical acts, Gurak and Antonijevic continue: "Speeches are recorded; sliced and diced and mixed; uploaded to Facebook and YouTube; streamed in chunks or snippets on news sites and blogs; and bounced on waves via broadcasts or satellites" (497). Furthermore, conversations "can rarely escape being digitized: Cell phones capture our winged words; hidden Webcams increasingly record our every public move" (497). In other words, for better or worse, the world as Ferris sees and represents it is the world that Gurak and Antonijevic theorize. It is a world in which individuals and communities alike must accept that the virtual and the so-called real exist in dynamic interplay with one another in and beyond the scope of the textual. It is a world in which, regardless of their religious or secular identities, individuals and communities alike might come to think somewhat like Betsy, who manages to see God "in the sky," "on the street," and even online, as evidenced by her remark that whether "it's online or offline, it's God's world" (Ferris, *To Rise Again* 33, 33, 217).

By reconciling with the reality of the hybridized world, twenty-first-century Americans as Ferris represents them have opportunities to discover approaches to making meaning in their lives that resemble the ways in which readers make meaning of multivalent texts both on- and offline. And they are invited to do so in the face of and perhaps to help counter digital-age problems such as those that involve alienation, identity, and authenticity. They have opportunities to find ways to avoid wasting their lives or at least the very real *feeling* that they are wasting them, to reference Paul's concern that he has "wasted [his] life" (8). And they have opportunities to find ways to overcome the apperceived "glitch in the soul" that digital media engagement produces without disengaging entirely from a digital world that is always already part of physical reality (189). For instance, they can avoid the disjointed feel of modern life to which Ferris draws attention in "Fragments," a story that in literary form makes an argument akin to Turkle's: that digital devices and media fragment modern life and that meaningful conversation as a form of authentic connection might function as the "cure" (*Reclaiming Conversation* 5). In the short story, Ferris focuses on the life experience of a protagonist who comes to feel a sense of disconnection and fragmentation that is analogous to Paul's, though he feels disconnected for a notably different reason. After receiving an accidental cell phone call from his wife, Katy, and overhearing a bit of the conversation she attempts to have in private with her lover about spending the night together, the protagonist of "Fragments" experiences a challenge to his blind faith in his wife, coming to the realization that his wife is having an affair. Over the course of the following hours and days, he experiences in a different context the "fragments" that "came through," first "[a]mplified" and "then muted," in the accidental phone call from his wife ("Fragments" 160). As he moves through his city's streets, he hears bits and pieces of cell phone conversations and in-person conversations that sound as though they are cell phone conversations. In other words, he hears and sees that the sense of fragmentation which manifests within him as a result of the collapse of his marriage has, for better or worse, a counterpart in the nature of everyday modern existence in part as a result of the pervasive reach of digitization.

In particular, Ferris insinuates that contemporary Americans such as the unnamed protagonist of "Fragments" or Paul from *To Rise Again at a Decent Hour* might find some semblance of meaning in life much as readers find meaning in texts: they could establish a healthy degree of skepticism that runs counter to the fundamentalist "[l]iteral doubt" of Ulmism and that also opposes blind faith, which characterizes the unnamed protagonist of "Fragments" (*To Rise Again* 162). Through healthy skepticism, they can function as effective critical thinkers in the globalized world and effectual, multiliterate critical readers of digital texts and aspects of that hybridized world as a text: as the New London Group theorizes it, multiliteracy is a phenomenon that accounts "for the context of our culturally and linguistically diverse and increasingly globalized societies" and "the burgeoning variety of text forms associated with information and multimedia technologies" (61). In addition, they can emerge as better readers of more conventional texts such as Ferris's novel, which presents them with different versions of Grant's personal history and the origin of Ulmism. According to a version that billionaire Ulmist Peter Mercer shares with Paul after finding him online through the web presence that Grant creates for him, Grant's history involves a more authentically religious calling. As Paul recounts Mercer's story, "Over time, Arthur made one discovery, and then another, and another and another about his ancestry, and about who he really was" and feels "duty-bound to leave" his Orthodox Jewish love, Mirav Mendelsohn, Rabbi Osher Mendelsohn's daughter, to reestablish "a community of diasporic Ulms" (280). According to another version of Grant's history that Connie's uncle Stuart unearths on the internet, Grant has an opaque and seedy past. He comes from a wealthy New York family, moves to Los Angeles, changes his name to David Oded Goldberg, and gets arrested for "harassing an Orthodox Jewish rabbi named Osher Mendelsohn," who took out a restraining order against Grant for reasons not specified online (249). According to yet another version of Grant's story told by Mirav, who is found by Stuart and Mercer's private detective, Wendy Chu, in an Orthodox Jewish "campus or housing network," there is "something desperately fraudulent" about Grant, though the nature of his fraudulence is nebulous (281, 304). She explains that he had sought to convert to Judaism despite his faithlessness because he loved her. But she and her father came to question Grant's sanity and intentions. As a result, she leaves both Judaism and Grant. As Stuart puts it, Mirav's story suggests that Grant is a "man broke with reality. He took an old legend from the Bible and made a myth from it, and now he tells the myth like it's truth" (307).

Ferris intimates that the unknowable, elusive, and subjective truth of Grant's history is less important in the digital age of misinformation than the ways in which characters and readers alike approach having increasingly more access to different untruths, truths, and versions of stories, including Grant's story, because access makes critical thinking, reading, and meaning-making more necessary. Certainly, Paul struggles deeply to negotiate with variations of Grant's story and divergent accounts of the origins of Ulmism. In a moment of despair, he altogether avoids addressing the variations he encounters by getting drunk on his balcony overlooking the bustling Brooklyn Promenade, blacking out from alcohol consumption, and waking up at an indecent hour to solitude that makes him feel his life lacks "meaning" (310). Yet he distinguishes himself by way of his ability to move beyond the chaos of divergence. He is able to transcend the problem of different possible realities, imagining, in a line that alludes to the title of Ferris's novel, the possibility of rising "again at a decent hour" and moving forward to make meaning in his life (309). By contrast, as a result of hearing Mirav's account of Grant's life, Mercer "bought a gun, walked out into the woods, and shot himself in the head" (314). He shoots himself in a weak act that echoes that of Paul's father, presumably because Mirav's account destabilizes the foundational narrative on which he bases his identity. And he shoots himself presumably because he feels unable to remedy his fatal flaw—his credulity—which Mirav's story illuminates for him. He presumably feels unable to move beyond the sort of blind faith that initially fuels his devotion to evangelical Christianity and that eventually fuels his devotion to doubt. And he likely is unable to come to terms with the fact that his devotion to doubt manifests as a result of a head injury on a C-train platform (261). Finally, he is incapable of critical thinking that would lead him to come to terms with the fact that Grant may well have targeted him as a convert to Ulmism because of his wealth.

Furthermore, Ferris suggests that work and perhaps, too, a Protestant work ethic, which Max Weber identifies as essential to capitalism and which Americans enact in near-fundamentalist ways,[8] can function to give meaning and meaningful community to otherwise alienated twenty-first-century citizens—even if the work results in the production of art that has varying degrees of social influence. Reflecting on work in his discussion of Paul's profession in his interview with Bollen, Ferris observes that "it would be very hard" for him "to create a novel in which a character's job was incidental to their identity" ("Joshua Ferris"). As Ferris explains to Bollen, his feelings about work emerge because his job is "so crucial" to his identity. "I wake up every day," Ferris continues, "in order to do something that's quixotic, and not necessarily called for in the world, but I do it because there's extraordinary meaning for me behind the effort. I'm not sure that I want to write a book about someone who doesn't take life seriously in that manner." As a result, he says to Bollen, his "books are about work." Also, his books aim to do important work toward the social good. As Ferris remarks to Bollen, writers are "generally charitable people. They're interested in the world." They mirror, for instance, the kind of social concern for suffering that Marc Chagall shows in *Solitude*, a painting that Grant owns in its apparently original form—if Grant and the seller of the painting are to be trusted. However, Grant likely fails to understand the socially concerned work that Chagall aims to perform by virtue of portraying a Jew who laments the persecution of Jews and by showcasing what David Lyle Jeffrey calls "a fusion of symbolic meaning" that achieves "a simultaneous celebration and integrative reinterpretation of both Jewish and Christian sources of consolation" (213). In other words, Grant's apparent illiteracy toward the painting's significance illustrates Ferris's view that art may be futile as an agent of change despite the intentions that its creators have. As Ferris tells Bollen, what writers do "is essentially useless. Except for the sake of the thing itself. It's an interesting dynamic that this over-riding preoccupation to get back to the page also happens to coincide with our continued irrelevance in the world. It's a great and frustrating contradiction. Because I think writers do want to make an impact on the world."

In the world of his fiction and for an audience of readers who ideally approach his novel with an investment in thinking critically about subjects such as work, faith, and suffering, Ferris explores possibilities for unconventionally artistic, meaningful, and practical work in the world that might complement the work of artistic or literary production. As the novel draws to a close, Ferris portrays Paul as no longer present without presence at his dental office. Instead, Paul is more mature and committed to creating a bridge between the aesthetic concerns that occupy Ferris's imagination and social concerns involving health and wellness. Paul continues to create whiter and straighter smiles, but he does more than that. He thinks critically about his patients' problems much as Ferris wants his readers to think critically about his text and the world. And he shows an investment in humanity, as evidenced by his interaction with a pregnant patient who might have a cavity. After examining her, Paul surprises himself by telling her that she has "nothing to worry about right now" (319). He surprises himself because he exhibits a value for not "dwelling on all the shit and misery" (319). He values her desire to keep her unborn child healthy and to retain mental health herself during her pregnancy. And he feels the experience is a revelation, observing, "I was on the inside with this thought. No longer alien to the in, but in the in" (319). He even realizes, as a result of the experience, that he seeks to rekindle his romance with Connie.

Ferris likewise suggests that professions such as Paul's allow workers to perform meaningful work toward the social good in the nondigital world in ways that complement the work that authors engage in and dovetail with social work that American religious believers may perform through organizations such as the Salvation Army or Habitat for Humanity. And he suggests that this kind of social work allows individuals to transcend fixations on paradoxes of truth and knowing. They can focus on more concrete actions they can take to make the world better. After Connie turns down Paul's marriage proposal, quits her job as his office manager, and reveals her plan to move to Philadelphia with her new partner, Paul combats his depression by working for a humanitarian cause in his community. He develops a friendship with and treats for free his new patient Eddie, a depressed octogenarian. In turn, in the novel's epilogue, Paul becomes a humanitarian beyond his local community, as demonstrated by his newfound perspective on overseas clinical work. At the novel's start, Ferris portrays Paul as going to New Delhi with Betsy "for tax reasons" and "roasted lamb" (26). But by novel's end, after "tending to the teeth of the poor and malnourished," Paul photographs poverty as he encounters it to showcase that there exists an interplay between the sleek digital world and the textured and tragic reality (334). He seemingly comes to inhabit the real world more fully and sees it through more compassionate eyes.

Moreover, Ferris underscores ways in which humanitarian work can catalyze emotional and psychological transformations that help twenty-first-century Americans find meaning in digital-age life through nonreligious means. In the novel's epilogue, Ferris represents Paul as overcoming the trauma of his father's suicide while he is working abroad by abandoning his unhealthy Red Sox fanaticism, which emerged out of Paul's relationship with his father. Ferris likewise portrays Paul as coming to terms with the notion that his dedication to baseball should involve more than mere dedication to a team. To quote the words of Paul's friend McGowan, Paul realizes that "engaging with the world" involves more than "watching a Red Sox game" on a television screen, an antecedent to digital screens that pervade twenty-first-century life (263). By novel's end, Paul appreciates baseball for its capacity to teach him perseverance. He dedicates himself not to the Red Sox, a team that defines his past, but to the Cubs. He supports a team that is committed to a process of working to succeed against all odds to end their drought and win the World Series as the 2004 Red Sox won it. In Paul's religiously infused words, "Imagine it! Joining in the preseason to pray for a good year, watching their performance with genuine suspense, and feeling again the crushing heartbreak that only the perennial, tantalizing possibility of true redemption can provoke. My God! The world was new again!" (335–336).

Finally, Ferris draws attention to ways in which meaningful humanitarian work can create community across national borders, creeds, races, ethnicities, and linguistic heritages in a digital age of passivity and disconnection. In this period in history, to appropriate Paul's reflection on his increasing sense of disconnection from Connie, "what separated the living from one another could be as impenetrable as whatever barrier separated the living from the dead" (318). Yet Ferris concludes his novel with an image of action and connection: Paul playing cricket with a Nepali kid. Taking a plank from the boy, Paul plays a game that he does not understand. As Paul explains, "without any expectation or understanding, doubtful of any hope of success, I swung, one eye on the ball, and one eye on heaven" (337). Accordingly, by novel's end, Paul accepts the certainty of uncertainty in his life despite notions that all information exists as readily available online with the click of a button. He comes to terms with his own unconventional, nonreligious notion of heaven and spirituality instead of attempting to adopt others' religious traditions. He comes to have a nonreligious faith in the value of playing the game, a metaphor for life in Ferris's novel.

Ultimately, Ferris's readers must also accept a sense of uncertainty in twenty-first-century America. Indeed, Ferris opts against presenting tidy answers to pressing questions, instead creating a messy and multivalent text that generates space for readers to feel lost and eventually come to different interpretations. Although Ferris's readers learn of the initial form of Paul's changed perspective on life, they fail to learn Paul's fate and the ways in which his perspectives develop. They only have an opportunity to value uncertainty and opportunity as forces that run counter to apparent digital-age certainties that only masquerade as true. They can contemplate ways in which they might learn from Paul's missteps and successes to reshape their own digital-age ways of thinking and being. And they can begin the process of imagining possible reconfigurations of their own identities and communities both online and off as Paul begins imagining them. Thus, Ferris reinvigorates his readers' imaginations as well as fiction as a medium with a social purpose. Although reading fiction or engaging with virtual reality certainly can never replace living life, and although fiction likely will not change the world, fiction can invite readers to puzzle over what it means to live in meaningful ways amid a proliferation of digital devices and media. It can invigorate human faith in a process of learning from the past to forge a meaningful digital future.

Judaism and the Digital Future in Jonathan Safran Foer's *Here I Am*: Other Lives and the Struggle for Ethical Presence in Twenty-First-Century Life

Through a hybrid form that incorporates digital messaging and by means of content that addresses human relationships with digital devices and with one another, Jonathan Safran Foer's *Here I Am* (2016), like Ferris's novel, represents the twenty-first century as media saturated. Foer's personal views on technology resemble Ferris's and inform his treatment of the digital age in his fiction, the focus of the second section of this chapter. In "Technology Is Diminishing Us," a 2016 *Guardian* essay, Foer outlines ways in which technology diminishes human experiences in the physical world. He acknowledges that "communication technologies began as substitutes" for otherwise impossible communication. Yet he critiques everyday users of digital devices and media for their preference for digital as opposed to non-digitally-mediated connection. He thereby comes into conversation with Jean Baudrillard's *Simulacra and Simulation*, a theoretical work that argues that people now lack the ability to distinguish between reality and simulation and that simulacra (simulations without originals) are ubiquitous in part because of the prominence of mass media. As Foer explains his critique, "we began to prefer the diminished substitutes. It's easier to make a phone call than to make the effort to see someone in person." Foer likewise suggests that the distracted state which digital media proliferate diminishes the human capacity for empathy and emotion. "The more distracted we become," according to Foer, "and the more emphasis we place on speed at the expense of depth—redefining 'text' from what fills the hundreds of pages of a novel, to a line of words and emoticons on a phone's screen—the less likely and able we are to care." Indeed, diminished human experiences in the world and diminished emotional registers may lead to the overall degeneration of humanity. As Foer asks, "Isn't it possible that technology, in the forms in which it has entered our everyday lives, has diminished us?"

In *Here I Am*, Foer most pointedly draws attention to the dystopian twenty-first-century reality that he sees digital devices and media as proliferating through his depictions of Other Life, a fictionalized platform that takes on a quasispiritual function in the novel, particularly for secularists. As Foer tells Terry Gross in a 2016 NPR interview, "I'm not a believer. I wish I were." And he observes to Gross that the agnostic Jacob resembles him in that he maintains "a religious identity largely through double negatives" ("Jonathan Safran Foer on Marriage"). According to Foer, Jacob and Julia Bloch "would do anything short of actually practicing Judaism to instill a sense of Jewish identity" in Sam and their other children. They practice what the narrator of *Here I Am* calls a "religion for two" (*Here I Am* 10). Perhaps as a result, Sam unearths a parody of transcendence in Other Life, a fictionalized virtual reality that is analogous to real-life platforms such as Minecraft or Second Life, both of which grew out of multi-user dungeons (MUDs), or text-based virtual worlds that offer users what Turkle calls "parallel identities, parallel lives" (*Life on the Screen* 14). Initially, much as Bill Peek of Zadie Smith's "Meet the President!" transcends the limits of physical life through his AG 12 device, Sam intentionally transcends the limits of his own body, social identity, and hence life by entering into Other Life via his Latina avatar, Samanta. Eventually, Sam enters into Other Life via Eyesick, an avatar that replaces Samanta after Jacob logs into Sam's account, takes Samanta "for a spin," and accidentally kills her by inhaling a Bouquet of Fatality (*Here I Am* 162). As Eyesick, Sam certainly transcends the limits of Samanta's virtual death. And he also transcends the material reality of his great-grandfather Isaac Bloch's suicide, which is the result of Isaac's feelings of abandonment by his family at the symbolic moment when a seemingly apocalyptic earthquake strikes Israel. In Eyesick, Sam gives Isaac a digitized afterlife that Judaism as a religion fails to provide for him definitively, as illustrated by what David S. Ariel calls a lack of a "consistent theory about life after death" in the Bible, particularly in the case of suicides (73). Sam also provides Isaac with an afterlife in which he is free from the great post-Holocaust threat to his life. He provides him with freedom from fear of displaced personhood: the threat of being forsaken by his family members and relocated to a "Jewish Home" for the elderly against his wishes (*Here I Am* 3).

Foer characterizes the transcendence that digital-media platforms such as Other Life purport to provide as detrimental for the ways in which they simultaneously connect and disconnect individuals from experiences in real life. As Foer intimates via his description of "The Harmony," the daily real-life moment at which natural light corresponds with the permanent dusk of Other Life, similarity between the real and the virtual is either mesmerizing or entirely unsettling (62). And unsettled feelings about the connection between reality and virtual reality likely develop because, in Turkle's words, "virtuality tends to skew our experience of the real in several ways" (*Life on the Screen* 236). Simulation, too, according to Turkle, makes "the fake seem more compelling than the real," as demonstrated in Foer's novel, for instance, by Other Life proponents who repeatedly insist that Other Life is "not a game" and as also demonstrated by the notions of community that Sam obtains as an Other Life user (*Life on the Screen* 237; *Here I Am* 163). Notably, Sam immerses himself in Other Life after Rabbi Singer accuses him of writing racial epithets on a note in Adas Israel Hebrew school. And Sam does so perhaps because the accusation threatens to thwart his bar mitzvah and thus his full entrance into the Jewish community, which bears an unlikely resemblance to Other Life in that it involves "a bunch of people" getting together to explore an "*imagined landscape*" for a monthly fee (161). Yet through Other Life, Sam fails to obtain a meaningful substitute for religious community in the physical world—even though real-life religious community for a child of agnostics in the United States may come with its own unique problems. Even if he fails to fully realize it, what Sam finds instead, to borrow Turkle's speculation, is that "virtual intimacy" may "degrade our experience of the other kind and, indeed, of all encounters, of any kind" (*Alone Together* 12).

The degradation of experience that Foer sees as manifesting in virtual reality tends to involve the devastation of conversation, which Turkle addresses in her book *Reclaiming Conversation*, arguing that face-to-face conversation is essential because it is "the most human—and humanizing—thing we do" (3). Foer best represents the problem of conversation in the digital age in a grossly disjointed textual exchange at Samanta's Other Life bat mitzvah. The exchange showcases the fragmented nature of digital-age life as both Ferris and Foer see it, and it complements post-traumatic fragmentation as Foer represents it in *Extremely Loud and Incredibly Close*.[9] As the pews in Samanta's virtual synagogue fill, numerous attendees, who are real-life strangers to Sam but familiar with Samanta, have a dialogue that Foer introduces in medias res about a mix of subjects including a phone, the purpose of Samanta's virtual gathering, and apparently ongoing real-life experiences that Other Life users are engaging in or perhaps largely ignoring while online. To excerpt Foer:

>Just take it to someone else. Insist that they open it up.
>Just fucking throw your phone from a bridge.
>Can someone explain to me what's going to happen here?
>Funnily enough, I'm crossing a bridge right now, but I'm on an Amtrak and you can't open the windows.
>Send us a picture of the water.
>Today Samanta becomes a woman.
>There's more than one way to open a window.
>She's having her period?
>Imagine thousands of phones washed up on the beach.
>Love letters in digital bottles.

> Why imagine? Go to India.
> Today she's becoming a Jewish woman.
> I'm on an Amtrak, too! (*Here I Am* 63)

Foer shows ways in which the fictionalized interlocutors struggle to follow along with various threads in the conversation, much as Paul in *To Rise Again at a Decent Hour* fails to follow what happens in his workplace because of digital distractions. And Foer intends for his readers to struggle to follow who is addressing whom in this excerpt as well. By virtue of this exchange, Foer invites his readers to question the degree to which advancements in technology confuse as opposed to improve American life. He invites them to question what Carroll Pursell identifies as a "belief" in the United States that technology holds "the key to a stronger, richer, healthier, and happier America" (xii).

Foer complements his critique of the degradation of dialogue at the hands of digitization with an interrogation of the effect digitization has on ethical sensibilities, which Lewis S. Gleich and Chris Vanderwees see as central to *Extremely Loud and Incredibly Close* and which religious believers of different kinds inevitably concern themselves with.[10] He suggests that the digital age creates new kinds of ethical dilemmas, a point that he makes poignantly via the narrator's narration of Jacob's thoughts about his and his family's reliance on digital devices and media. According to the narrator, "Ideally, we would [...] use some mental resource other than Google, and some physical resource other than Amazon, [...] and never put a child in front of a screen" (*Here I Am* 566). But, as the narrator continues from Jacob's perspective, "we live in the world, and in the world there's soccer practice, and speech therapy, and grocery shopping, and homework," among other obligations, "so as nice as that idea is, there's just no way we can make it happen" (566). Along the same lines, Foer suggests that users of digital platforms such as Other Life may feel ethical responsibility in life but may rebuff responsibility in virtual reality. In particular, Foer highlights the relationship between Sam's ethics in real life and his ethics online, which are problematic because Sam sees online environments as affording him freedom from real-life ethical constraints. According to the narrator's close third-person narration of Sam's thoughts while Sam as Eyesick trespasses in a virtual lemon grove in Other Life, Sam "would never trespass in life itself. He was too ethical, and too much of a coward. (Sometimes it was hard to differentiate.) But that was one of the many, many great things about Other Life—perhaps the explanation for his addiction to it: it was an opportunity to be a little less ethical, and a little less of a coward" (328).

Foer sees the kind of dissociation from reality that digital media facilitate as influencing violent behaviors as well. Earlier in the novel, Sam evokes Bill's cruelty and violent tendencies in Smith's "Meet the President!" when he engages in a far less ethical act than virtual trespassing. He performs an act of terrorism as Samanta online—even though he enacts it against his own virtual synagogue while no avatars occupy it. The narrator explains that he flattens "the synagogue to rubble," much as homegrown American terrorists flatten religious sites in real-life post-9/11 America (85). Sam's terrorist act prefigures Foer's fictionalized real-life terrorism, which takes place following the earthquake when "a squad of Israeli extremists penetrates the Dome of the Rock and sets it on fire" (281). Thus, Sam's act draws attention to the complex relationship between virtual and real violence. Even though, as Craig A. Anderson, Douglas A. Gentile, and Katherine E. Buckley write, research suggests that there exists a clear correlation between "media violence" and "increased aggression or violence," "unresolved issues" involving the relationship between media and real violence remain (4, 4, 5). Anderson, Gentile, and Buckley explore those unresolved issues and focus particularly on "violent video game effects" as researchers have heretofore not focused on them (5). They explore real acts of online violence such as the violence that Sam displays in fiction.

For Foer, digital devices and media, too, inhibit ethical behavior that religious believers concern themselves with in ways that degrade human relationships because these devices and media encourage the proliferation of metaphorical other lives that resemble Other Life in their function. For instance, Jacob lives another life much as his son Sam does through consistently listening to podcasts instead of engaging with his family. Most notably, he also lives another life through a second cell phone he purchases for the apparent sole purpose of sending sexually explicit text messages—or sexts—to one of his colleagues, who, like his wife, is named Julia (or is at least given the name of Julia and the status of a simulation or imitation by Jacob on his cell phone). Much as messages sent by users within Other Life come to comprise a peculiar and disjointed textual conversation, sexts sent and received by Jacob fragment Jacob's everyday life experiences, showcasing ways in which his other sexual life with his mistress devastates his home life by interweaving with and interrupting it. The first of Jacob's sexts that Foer's readers encounter appears immediately following a lengthy list of things that Jacob's wife, Julia, likes, and it precedes an equally lengthy list of things that his wife dislikes. It interrupts a narrative about mundane and monotonous domestic experience with bold, sexually explicit language that contains usage errors typical of texts generated in fast-paced contemporary life. And texts between Jacob and his mistress—the other Julia—continue throughout the chapter. They interrupt a narrative that describes, among other things, the notably disconnected sexual acts in which Jacob and his wife engage; the ways in which Jacob and his wife drift apart from each other; Jacob and Julia's effort to find happiness in their marriage by visiting a Pennsylvania inn that they had previously visited in the predigital age; and the dissolution of the marriage of Jacob and Julia's friends, Mark and Jennifer, which foreshadows the fallout between Jacob and his wife.

Foer intimates by virtue of his depiction of Jacob's unethical activity on his second cell phone that a key problem of living another life is that the other life which gets lived can never exist as wholly independent of real life, even though other lives certainly manage to confound notions of what counts as reality. In other words, Foer posits that even if engagement in other digital lives does not outright influence activity in real life, other lives in whatever forms they take—even if they manifest via identity theft as Paul experiences it in Ferris's *To Rise Again at a Decent Hour*—may haunt or damage the very real and quite fragile physical and emotional lives that digital-age Americans live. In the final text that readers see in the string of texts between Jacob and the other Julia, the other Julia abandons the language of sexual fantasy for pragmatic concerns and allows Jacob's other life to bleed into his real one. After calling and failing to reach him, other Julia writes, "*what happened to you?*" (78). And in writing this message, other Julia explicitly asks why Jacob failed to do his part to sustain their fantasy, which resembles the fantasy that the protagonists of Roupenian's "Cat Person" sustain through their texts. Other Julia interrogates the reason for the return of reality even though reality never vanishes as she might wish it to. And she unintentionally brings about in the Bloch household a revelation of infidelity and a real-life reckoning with it. She also brings about a reckoning with what comes to count as infidelity in the digital age and, to appropriate words from a conversation between Mark and Julia, whether "the distinction" between "talking and doing" and therefore between virtual and real-life cheating "matters" (170). Other Julia's phone call and her pragmatic text lead Jacob's wife, Julia, who is positioned as the real Julia, to discover Jacob's buzzing second cell phone on the bathroom floor behind the toilet. They lead her to discover that Sam previously dropped the phone there after being the first to find it. He had unlocked it with the thumb of his maimed hand, which was "crushed in the hinge" of a "heavy iron door" and which serves as a harsh reminder of the undeniability of physicality and reality, and he dropped it because he felt shocked by its contents (12). Other Julia's pragmatic text also prompts Foer's digital-age readers to reflect on what led Jacob to have the affair in the first place and whether it would have happened without the digital tool and culture that facilitated it. It leads readers to contemplate the dual nature of Jacob's identity, which Julia considers later in the novel when she tells Jacob that he is the only person she knows "who would be capable of writing such bold sentences while living so meekly" (117). Finally, the last text sustains a metafictional function. Through its second-person voice, it prompts readers to reflect on whether they, too, have changed, perhaps at the metaphorical altar of digitization.

Foer's portrayal of Jacob's cell phone as a gateway to a fantasy-laden other life underscores ways in which digitization fuels an American proclivity for desiring or outright living digital or nondigital other lives that may fragment existence. Hence it speaks to Don Ihde's argument in *Bodies in Technology* that there are nontechnological means by which to experience virtual reality, which Ihde sees as creating anxiety among the masses who fear that it will "supplant or replace" real life (4–5, 3). For instance, after the trauma of hiding in "a hole for so many days" that "his knees would never wholly unbend; among Gypsies and partisans and half-decent Poles; in transit, refugee, and displaced persons camps" and in other locales as part of his escape from the Nazis during World War II, Isaac, like other members of the Jewish diaspora and like refugees in Hamid's *Exit West*, desires and seeks out a virtual reality in the form of another life (Foer, *Here I Am* 3). He settles in the greater Washington, DC, area, and raises Jacob, who, with Julia, raises Sam, Max, and Benjy. His brother Benny settles in Israel and raises Jacob's cousin Tamir, who, with his wife, Rivka, raises Sam, Max, and Benjy's cousin Noam. The two branches of the same family live lives that are other to one another, so to speak. The former faces problems involving capitalist American modernity, and, particularly following the earthquake in Israel, the latter faces religious conflict in everyday ways that are evocative of the Tanakh's portrayal of religious clashes. Since Tamir is visiting with Jacob during the earthquake, Tamir's cell phone helps him maintain contact or a sense of closeness with his family, and he consequently turns to it in markedly different ways than Jacob turns to his. Despite his career in tech, he escapes what Jacob terms "the Great Flatness," presumably a disconnected state of being that metaphorically goes viral due to the ubiquity of flat screens (392). Tamir is also forced to live a different life than the one he feels a social responsibility to live because of the earthquake and the war. He must remain separate from his family and his son Noam, who fights in the Israeli army when war breaks out. As evidenced by his decision to transfer his resilience fruit to Eyesick, Sam's Other Life avatar, in Other Life (in order to give Sam another life within Other Life), he fears that war will take his life and lead him to yet another life: life after death as many Jews believe in it. Jacob, too, contemplates living another life: he considers the option of traveling to Israel to fight in the army after the Israeli prime minister makes a Zionist call for Jews to "come home" as part of "Operation Arms of Moses" (436, 540). Even the model United Nations field trip that Foer portrays functions as a simulation and another life. It enables Sam; Billie, Sam's love interest; and Sam and Billie's classmates to address Micronesia's theft of a nuclear weapon, the scenario with which the simulation's leaders opt to present students. And it also enables Julia to flirt and pursue an affair and another life with Mark. It allows her to play out a romantic fantasy that is similar to her sexual and other fantasies, for instance the fantasy she has of her death, which she has just prior to hearing Jacob's buzzing second cell phone and finding it. As Foer depicts Julia's movement between fantasy and reality and between internal and real lives, she is brought "back to life" by "a buzzing—it shook her free from her unchosen fantasy, and she was hit by the full absurdity of what she was doing. Who did she think she was? Her in-laws downstairs, her son down the hall, her IRA bigger than her savings account" (77).

Perhaps most notably for Foer as a writer, fiction as a medium provides an opportunity for both writers and readers to live other lives. These other lives formerly existed in nondigital environments, but they increasingly exist in the digital world since, as N. Katherine Hayles argues in "The Future of Literature," "print and electronic textuality deeply interpenetrate one another" (181). According to Hayles, despite their different functions, electronic and print texts are "two components of a complex and dynamic media ecology" because "they engage in a wide variety of relationships, including competition, cooperation, mimicry, symbiosis, and parasitism" ("The Future of Literature" 181). In other words, these texts help to construct and also reflect the reality of hybridity as a key feature of twenty-first-century life. Moreover, these texts and arguably all works of fiction reflect the hybrid interplay of truth and lies, as evidenced in *Here I Am* by the interplay of Foer's life with Jacob's, which functions as an other life for Foer, and by the interplay of Jacob's life with the other life of Jacob the television show character, the protagonist of *Ever-Dying People*, the HBO miniseries that the fictionalized real Jacob writes. Foer's life resembles the life of the real Jacob of *Here I Am* in a number of ways, much as the fictionalized Jonathan Safran Foer's life in *Everything Is Illuminated* resembles the real Jonathan Safran Foer's life. Both Jacobs hail from Washington, DC. Both have written television shows about secular Jewish families.[11] Both won the National Jewish Book Award at the age of twenty-four.[12] And both write about their life experiences in veiled ways, as best evidenced by the fact that Foer and Nicole Krauss divorced prior to the publication of *Here I Am* just as Jacob and Julia divorce within the world of the novel. Although Foer observes in a 2016 *Independent* interview with Heller that "[t]here's nothing to recognise" in the novel, he tells Terry Gross, "I sometimes can't remember what's in my books and what's in my life" ("Jonathan Safran Foer Interview"; "Jonathan Safran Foer on Marriage"). Perhaps like Jacob, whose "writing kept pace with the changing events" of his life or whose "life kept pace with the writing," Foer lives a life informed by his fiction and does not just live a life as a writer who writes about life in his fiction (*Here I Am* 200). Hence the truth of Foer's novel likely resembles the truth of Jacob's television show. In Jacob's words, should he one day "be asked how autobiographical" his show is, "he would say, 'It's not my life, but it's me'" (199). As a result, Foer suggests, to borrow a transliteration that Sam shares with Billie on two occasions in the novel, "Emet hi hasheker hatov beyoter," or "Truth is the safest lie" (154).

Foer positions the bible for Jacob's show, a text that serves as a "user's manual for those who would one day work on" *Ever-Dying People*, as illustrative of the relationship between Jacob's fictional real life and his fictional other life within the world of Foer's fiction as the character of Jacob in *Ever-Dying People* (200). In doing so, Foer comments on the Hebrew Bible much as Ferris comments on the Christian Bible in *To Rise Again at a Decent Hour*. Foer delineates the relationship between the Hebrew Bible, after which Jacob's bible is modeled, and real life as Foer understands it. He suggests that the Tanakh serves as a guide to life, particularly digital-age life that involves a proliferation of other lives, specifically because it interweaves fact and fiction. As Coogan explains, the Old Testament's (or the Tanakh's) "narrative framework" was, until the Enlightenment, "considered historical in the sense that it was accepted as an accurate, even inspired, account of what had taken place over thousands of years" (23). But, as Coogan expresses it, the Old Testament (or the Tanakh) "is also imbued with myth" (33). According to Coogan, "[m]yth and history" relate because history has "a mythical dimension" and myth has "a historical dimension" (39). And the "farther back we go in the biblical narrative," for instance to Genesis as Foer references it, "the more we are in the realm not of history but of myth" (Coogan 32). Indeed, the digital age in which Jacob writes his bible in many ways reflects the interplay of fact and fiction that these older narratives in the Tanakh in particular represent. To reference Keyes's term from his 2004 book, *The Post-Truth Era: Dishonesty and Deception in Contemporary Life*, digital times exist as post-truth times because the "World Wide Web is a mishmash of rumor passing as fact, press releases posted as news articles, deceptive advertising, malicious rumors, and outright scams" (205). Using the Tanakh as a model, Jacob merges fiction and history in *Ever-Dying People* and in his bible for the show, much as web users merge them in Ferris's *To Rise Again at a Decent Hour*. Jacob merges fact and fiction, which readers of Foer's novel may read as indicative of real-life events in Jacob's life in Foer's fiction, even though his show may only function as fiction and even though his bible may well function as an information-dense guide to reading fiction for citizens of an increasingly post-text and information-dense future. It may function as a road map to literature in an era in which literature exists as the stuff of another life, a notion that Sven Birkerts spotlights in *The Gutenberg Elegies: The Fate of Reading in an Electronic Age*, which prophesies the death of literature.

Foer appears to show interest in the ubiquity of other lives that global citizens of the digital age live because those other lives create an unsettling sense of absence—a sense that, in Hamid's words from *Exit West*, individuals are constantly "present without presence" because of the digital devices they own and the effects that those devices have on their desires and identities (40). For Foer, this digital condition of absence that defines the times stands in stark contrast to complete presence of the sort that his novel's title speaks to by referencing language from Genesis 22, a passage from the Tanakh that Foer perhaps engages with to reflect his deepening relationship with Judaism.[13] In Genesis 22, God tests Abraham's faith by asking him to sacrifice Isaac, his son and the biblical namesake for the Bloch family's patriarch. Upon initially hearing God's call, Abraham answers *hineni*, Hebrew for "here I am," a word that Foer, in an interview with Gross, says "has a real sort of presence in the Jewish consciousness" (*Tanakh*, Genesis 22:1; "Jonathan Safran Foer on Marriage"). Abraham then utters the same phrase in response to Isaac when Isaac, failing to recognize that God has called Abraham to sacrifice him, addresses Abraham while walking up Mount Moriah to inquire about the absent sheep for the offering that he believes he and Abraham must make (Genesis 22:7). Then Abraham utters the phrase a third time when an angel of the Lord calls to him to stop him from slaying Isaac on the altar and offers a ram to him to slay as his offering to God (Genesis 22:11). As Foer sees it, according to remarks he makes in a 2016 Goodreads interview with Jade Chang, these answers of *hineni* in the biblical chapter function as a "paradox" ("Interview with Jonathan Safran Foer"). In Foer's words, "How can [Abraham] be unconditionally present for God who wants him to kill his son and for his son?" ("Interview with Jonathan Safran Foer"). To build on Foer's remark, how can this representation of unconditional presence help contemporary readers of the Tanakh or of Foer's novel resolve the contemporary problem of absence?

By representing Sam's bar mitzvah speech, which addresses Genesis 22 in explicit ways in its draft and in implicit ways in its final, apparently impromptu, delivered version, Foer invites his readers to deepen their understanding of unconditional presence as a potential means by which to live a more meaningful life amid digital-age impulses for absence. Readers encounter Sam's explicit meditation on Genesis 22 in the draft of his bar mitzvah speech, which he delivers as Samanta online in Other Life. In the speech, Sam as Samanta argues that although most "people assume that the test" of Genesis 22 involves God's call to Abraham to "sacrifice his son," the test actually involves the initial call itself, and Abraham passes it by being "wholly present" and not saying "What do you want?" or "Yes?" (Foer, *Here I Am* 102). Via this reflection on Genesis 22, Sam as Samanta concludes that the biblical chapter "is primarily about who we are wholly there for, and how that, more than anything else, defines our identity" (103). Sam as Samanta concludes that Jacob and Julia failed to be present for Sam when Rabbi Singer accused him of using profanity. In Sam's words as Samanta, "I wish I had been given the benefit of the doubt," not because "I'm a good person," but because "I'm their child" (103). By contrast, Sam delivers a very different, apparently impromptu version of his speech in real life at the bar mitzvah that ends up taking place in the kitchen of his home because, as Sam observes, his original speech "wouldn't make any sense now, given that everything has completely changed" (450). Readers read Sam's speech over the course of several chapters, and they are interrupted by chapters that present speeches on the crisis in Israel given by the ayatollah in Tehran and the Israeli prime minister in Jerusalem. These interrupting chapters give Sam the status of a religious or national leader as he speaks on the subject of a personal (not national or international) crisis. As the narrator explains, Sam begins his speech with "the grace of being fully present" (449). And in his speech, he gives the subject of complete presence a notably absent presence by making no explicit mention of Genesis 22.

Instead, Foer has Sam focus his impromptu speech on the subject of choice, and in doing so, he illuminates the connection between complete presence and free will that, according to Ariel, "rabbis believed that humans possess"—even though there are consequences for "every action" and even though "some sages held that God has prior knowledge of what action will occur" (96, 96, 97). Foer suggests that choices pervade contemporary existence, as evidenced by the Yiddish phrase "[k]ein briere iz oich a breire," or "[n]ot to have a choice is also a choice," which the rabbi who speaks at Isaac's funeral describes Isaac as having taught him prior to his death (*Here I Am* 347). Foer suggests that the desire for complete presence as opposed to presence without presence inevitably leads mature individuals to have to make hard choices. As Foer intimates in his Goodreads interview with Chang, choice functions as the inevitable end to the story of Abraham's complete presence for God and Isaac in that Abraham must choose between them. As Foer explains in reference to the paradox of Abraham's complete presence for God and Isaac, "I really love those paradoxes of identity, like being a father and an ambitious professional; maybe having a firm political stance while having a certain religious stance; maybe being in a marriage while also having certain ideas of selfhood in the world" ("Interview with Jonathan Safran Foer"). "These are things," he continues, "that normally you can push to the back burner and they don't cause any real destruction or pain, but sometimes a crisis will force a choice—the crisis of a discovered phone, the crisis of an earthquake—and suddenly these things that we've been keeping distant or keeping dark become very alive and bright" ("Interview with Jonathan Safran Foer"). Sam appears to agree with Foer. For Sam, who observes, "I did not ask to be a man, and I do not want to be a man, and I refuse to be a man," choice exists as the paradoxical and undesirable stuff of adulthood, an enigmatic state that Abraham embodies from Sam's perspective (*Here I Am* 452). As Sam explains, exhibiting the ways in which the story of Abraham's ability to sustain complete presence confuses him, "You can't stop things from happening. You can only choose to be there, like Great-Grandpa Isaac did, or give yourself completely over, like my dad, who made his big decision to go to Israel to fight. Or maybe it's Dad who is choosing not to be there, which is *here*, and Great-Grandpa who gave himself over completely" (451–452). As Sam elaborates, showcasing his desire to avoid hard choices while referencing Billie's reflections on Shakespeare's *Hamlet*, "maybe one doesn't have to exactly choose. 'To be or not to be. That is the question.' To be *and* not to be. That is the answer"—at least for someone who wants to avoid manhood or adulthood in general (452).

Foer suggests that having faith in the value of making hard choices and actually making them gives individuals power and helps them lead more responsible lives in a digital age that is fraught with paradox. As evidenced by choices that Sam makes in giving his bar mitzvah speech, adulthood seems within reach for Sam— even though it remains undesirable in his view. For instance, Sam makes quite an adult choice to tell the truth about his actions in Adas Israel Hebrew school. He reveals that he is guilty of writing the racial epithets that Rabbi Singer accused him of writing. As he remarks in his impromptu speech, referencing the terms he opted to write, "I was just seeing how each of them felt, seeing how hard it was to write them, and say them to myself. That's why I did it" (457). Sam even makes the adult choice to explain that he sees that writing the words was "a mistake" (457). And he realizes that the contexts for adult choices make adult choices hard. In Sam's words, the "the hardest thing to say isn't a word, or a sentence, but an event. The hardest thing to say couldn't be something you say to yourself. It requires the hardest person, or people, to say it to" (457). Adulthood likewise seems within reach for Sam even though he sees hard choices as potentially violent in his childhood imagination, as evidenced by the association he makes in his speech between expressing a choice and using a nuclear weapon of the kind that Micronesia steals in the Model UN scenario. As Sam puts it, the people have nuclear weapons "to never have to use them," and contextually, readers infer that Sam believes that people have power to make choices so they never have to make them, particularly when choices they face resemble the one he feels he faces between his parents because of their looming divorce—a choice he suggests he can make (462). As Sam explains, if he were "forced to choose" between his parents, he "would be able to" (462).

For Foer, the thorny process of making hard choices appears to intertwine with the meaning of Israel, and it thereby exists as part and parcel of religious identity, which Foer conceives of in broad terms according to remarks he makes to Gross. Referencing Abraham Joshua Heschel's *Man Is Not Alone: A Philosophy of Religion*, Foer argues that "everybody is religious" ("Jonathan Safran Foer on Marriage"). As Foer elaborates to Gross, "we've just been sort of too vigilant about our terminology and our definitions" ("Jonathan Safran Foer on Marriage"). Foer showcases his conception of the far reach of religious thinking and experience through a meditation on Genesis 32, the subject of the bar mitzvah speech that Max delivers in the bible for *Ever-Dying People* (in "The Bible" chapter of Foer's novel). In Genesis 32, which Foer sees as broadly relevant beyond the bounds of Judaism, the biblical Jacob, a trickster who steals his brother Isaac's blessing from their father, Esau, and who serves as the namesake of Foer's fictional protagonist, "wrestle[s] with" a man who might be an angel or God himself "until the break of dawn" (*Tanakh*, Genesis 32:25). He holds him until the man agrees to bless him. The man then blesses him with the name of "Israel" since Jacob has, in accord with the meaning of Yisrael or Israel, "striven with beings divine and human" or, more simply, struggled or wrestled with them, and "prevailed" (Genesis 32:29). The passage thus makes apparent not only the character of Jacob, who is renamed Israel, but the character of Israel as a nation. As the fictionalized Max of Foer's bible puts it, "*Israel*, the historical Jewish homeland, literally means 'wrestles God.' Not 'praises God,' or 'reveres God,' or 'loves God,' not even 'obeys God.' In fact, it is the *opposite* of 'obeys God.' Wrestling is not only our condition, it is our identity, our name" (*Here I Am* 511). And as Max elaborates in his speech, wrestling is important because it produces "closeness"— for anyone, Foer seems to suggest, and not just for Jewish people (511). According to the fictionalized Max of Jacob's bible, "It's easy to *be* close, but almost impossible to *stay* close" (512). And, for this fictionalized Max, wrestling "can keep something close over time" because what "we don't wrestle we let go of. Love isn't the absence of struggle. Love *is* struggle" (512). Consequently, for Foer, anyone who seeks to enact love in life must engage in a struggle, and anyone who engages in a struggle to love is religious.

As Foer sees it, wrestling not only defines and sustains love, but it characterizes or perhaps even produces maturity, a theme that Ferris in *To Rise Again at a Decent Hour* and Shteyngart in *Super Sad True Love Story* show interest in. Wrestling produces maturity in a digital age that allows users of technology to opt into other lives instead of struggling with challenges in their own lives in accord with the meaning of Israel. Arguably the most significant act of wrestling that Foer represents in his novel is not the divorce of Jacob and Julia, who essentially opt out of wrestling with each other, but an act that involves Jacob's love for his elderly dog, Argus. In short, Jacob wrestles with the reality of Argus's declining health and with the prospect of euthanizing him, instead preferring to cling to him much as the Israelites clung to the broken tablets described in *Ki Tissa*, Hebrew for *when you take*, a portion of Exodus involving Moses and the tablets on which the Ten Commandments were written and the part of the Tanakh that Jacob addresses in his own bar mitzvah speech. As Jacob explains, he remembers his chosen passage mostly due to its connection to a related passage in the Talmud, rabbinical writings that present Jewish law and theology through interpretations of narratives from the Tanakh. This piece of Talmudic text that Jacob recalls introduces a question involving the broken tablets' fate: it asks, as Jacob's rabbi puts it, "Why didn't [the Israelites] just bury" the broken tablets, "as would befit a sacred text?" (445). Instead, because "God instructed Moses to put both the intact tablets and the broken tablets in the ark," the "Jews carried them—the broken and the whole—for their forty years of wandering, and placed them both in the Temple in Jerusalem" (471–472, 472). As Jacob reflects long after his bar mitzvah but at a point at which he still behaves in many ways like a child, as evidenced by his unwillingness to wrestle through problems in his marriage with Julia, the Israelites avoided burying them because "they were ours" (472). In other words, he sees the narrative of the tablets as expressing the value of ownership—a point with which rabbis analyzing the narrative may certainly agree. However, he fails to see that value and maturity also exist in the process of letting beloved objects, individuals, and creatures such as Argus go, as demonstrated by his reluctance to acknowledge the reality of Argus's suffering during a visit to Dr. Shelling, Argus's veterinarian. He fails to embody fatherhood and adulthood because he does not see, to reference his father's words to him in his youth after his father disposes of an ant-covered dead squirrel in front of their house, that *"When you're a dad, there's no one above you. If you don't do something that has to be done, who is going to do it?"* (357).

Foer intimates that letting go as opposed to maintaining closeness can paradoxically be indicative of closeness as well as maturity in a digital age that for him is in many ways defined by paradox. Notably, by the end of Foer's novel, Jacob finds maturity much as Paul finds maturity and meaning in life in *To Rise Again at a Decent Hour*. Jacob overcomes his feeling that he "couldn't do it" and that "Argus wasn't ready" (435). Even though he opts against acting on his impulse to leave his own real life in the United States for another life of fighting for Israel in the crisis that follows the earthquake—an impulse that functions as the catalyst for Julia's request that he put Argus down because *"it's time, and because he's yours"*—he complies with Julia's request because it comes to function as "an open, if invisible, wound" following Jacob's decision to return home from the Islip airport (429, 552). By declaring "I'm ready" to the vet in his final words of the novel, he overcomes the feeling of unpreparedness (571). And he overcomes the "feeling of not wanting to live in the world, even if it was the only place to live" (569). Thus, he comes of age in a way that perhaps all Jewish children fail to come of age at their bar or bat mitzvahs. And his action enables Foer to define coming of age as essentially coming to terms with reality and engaging in mending in the face of digital-age fragmentation. As Foer articulates to Gross, *Here I Am* is ultimately "a book about people trying to mend things, even at the expense of acknowledging an end when necessary" ("Jonathan Safran Foer on Marriage"). By engaging in acts of mending, users of new media can avoid fragmentation and have more integrated human experiences. As Foer explains to Gross, "Here I Am" means "live your life," a phrase that marks the end of a Maurice Sendak interview which Foer admires ("Jonathan Safran Foer on Marriage"). In Foer's words to Gross, living a whole life is precisely "what Jacob and Julia and Sam and the others are wrestling with, how not simply to move through days, but to move through days as oneself, as an integrated person" ("Jonathan Safran Foer on Marriage").

Foer indicates that to live whole as opposed to fragmented lives, twenty-first-century Americans and citizens of the globe must realize that wholes are comprised of parts much as the Wailing Wall to which Foer continually makes reference in his novel is comprised of stones and "folded prayers" pressed between them in messy ways (*Here I Am* 13). Americans and citizens of the globe must wrestle with the place of digital devices and media as parts of the world. They are parts that literary texts such as Foer's can help them understand much as bibles both religious and secular can help clarify the script that is life. As Foer writes in his 2016 essay in the *Guardian*, countering arguments such as Philippe Codde's, which suggest that Foer is most interested in image culture as the digital age fuels it,[14] "The novel has never stood in such stark opposition to the culture that surrounds it. A book is the opposite of Facebook: it requires us to be less connected. It is the opposite of Google: not only inefficient, but at its best, useless." In the words of the rabbi who speaks at Isaac's funeral, novels are important for their words because words as Jews understand them are "generative," a point that the rabbi illustrates by citing the oft-cited phrase from Genesis, "Let there be light," which results in the production of light and which helps illustrate Jacob's and Foer's conception of the relationship between religion and writing (*Here I Am* 350). As Jacob expresses it when he returns home to write his bible for *Ever-Dying People* following the trying experience of attending Julia's wedding to another man, "My synagogue is made of words" (533). As he concludes his rumination on writing, seemingly reflecting Foer's own perspective on the very real religious experience of living life in and through books and words as a writer, "I was inside the Holiest of Holies all along" (533). In other words, through writing, both Jacob and Foer metaphorically inhabit the most sacred area of the Temple in Jerusalem—the area that contained the Ark of the Covenant, a symbol of God's relationship with Israel. Thus, through writing, Foer and his literary alter ego sustain a relationship with God.

Yet, ultimately, Foer complicates his dualistic representation of the interconnected digital age and the notion that books alone might offer salvation to digital-age humanity that appears to be succumbing to the so-called "Great Flatness"

(392). As Foer concludes in his 2016 *Guardian* essay, "It's not an either/or situation—being 'anti-technology' is perhaps the only thing more foolish than being unquestionably 'pro-technology'—but a question of balance that our lives hang upon" ("Technology Is Diminishing Us"). Hence the novel as a medium that stands in stark contrast to digital media offers not only an other life to those media, a world of fiction via which readers can escape the real world, but something far more valuable. It offers a means by which to commemorate aspects of the pre-digital past, as evidenced by the effective way in which it commemorates other disappearing or lost aspects of life, most notably Isaac and all that Isaac represents as a Holocaust survivor. Indeed, Foer's *Here I Am* is an—or perhaps *the*—answer to the question that the rabbi at Isaac's funeral asks: the question of how we should "mourn Isaac Bloch," a question that the rabbi sees as having the capacity to invite an answer that might lead to salvation (353). Too, the novel offers an opportunity for "discovering the medium" of the book "anew," in Wurth's words about *Tree of Codes*, by presenting ideas for forging and navigating an integrated life. It provides, as Foer tells Chang, "a kind of funny, joyful, and occasionally tragic chorus of perspectives," and it "advances a lot of different ways of thinking about things—a lot of arguments" ("Interview with Jonathan Safran Foer"). In other words, the words of the novel as Ferris and Foer both see it are multivalent and generative of an array of philosophies, including philosophies about the kinds of relationships Americans have or might have with technology. The words of novels prompt readers to develop meaningful philosophies of technology of their own that help them find ethical ends and maturity. Without thoughtfully engaging with the perpetually relevant and important medium of the novel, they may risk a fate of absence in their own precarious present and ever-fragmenting twenty-first-century American moment.

CHAPTER 4

Cybercapitalism in Don DeLillo's *Cosmopolis* and Dave Eggers's *The Circle*

As illustrated by Thomas L. Friedman's appropriations of magical-realist rhetoric in *The World Is Flat*, the transcendent (as in otherworldly and apparently immaterial) elements of the digital age, which Joshua Ferris and Jonathan Safran Foer highlight through engagement with formal elements of religious faiths and sacred texts, come to shape notions of cybercapitalism, or the metaphorical marriage of digitization and capitalism. In Friedman's words, digitization "is that magic process by which words, music, data, films, files, and pictures are turned into bits and bytes—combinations of 1s and 0s—that can be manipulated on a computer screen, stored on a microprocessor, or transmitted over satellites and fiber-optic lines" (64). For Friedman, digitization is the driving force behind Globalization 3.0, a new era that is magically "shrinking the world from a size small to a size tiny and flattening the playing field at the same time" (10). Much as imperial forces initiated what Friedman terms "Globalization 1.0," and much as multinational corporations connected disparate elements of capitalist life to enable the "maturation of a global economy" in Globalization 2.0, the commercialized internet in the current, third phase of globalization has transformed the world (9, 10). In Robert W. McChesney's skeptical words in *Digital Disconnect*, it has "seemingly colonized and transformed everything in its path" (3). As McChesney argues, countering the rhetoric of enchantment of Friedman's book, "the Internet is changing capitalism in significant ways, and it may well assist those who wish to reform or replace it in the political arena; but it is not making capitalism become, in effect, for lack of a better term, a green, democratic socialist utopia" (15–16).

Through analyses of Don DeLillo's *Cosmopolis* and Dave Eggers's *The Circle*, this chapter tells the story of the nondenominationally transcendent and toxic effects of capitalism's hybridization, to reference Donna Haraway's and Homi K. Bhabha's theorizations of hybridity, which I characterize as mixtures and fusions that manifest in identity, in material and immaterial aspects of culture, and between aspects of identity and culture. This chapter also draws attention to ways in which literary fiction can do countercybercapitalist work. In the first part, I focus on DeLillo, who writes novels such as *Mao II* and *Underworld* that critics regularly describe as having prophetic qualities but who sustains a thorny relationship with the future he purportedly foretells largely because he is skeptical of the corporate-born digital technology that he sees as shaping it.[1] DeLillo, who is known for writing on a typewriter in the digital age (and who published *The Silence* in a font that makes it looks like it was produced on a typewriter), explains in a 2006 interview with Kevin Gray that the "nature of these devices is that we use them because we have them, not because they're necessary" ("Q&A"). Indeed, for DeLillo, technological devices exist as a byproduct of the capitalist system that he readily critiques throughout his oeuvre, most notably in *White Noise*, which depicts Hitler-studies professor Jack Gladney's absurd experiences in a consumer-oriented and screen-mediated America, and in *Underworld*, which considers the effects of the atomic bomb and the internet on twentieth-century American life. Certainly for DeLillo, technology as American corporations produce it functions as "our fate, our truth," to quote his meditation on technology and its connection to the 9/11 terrorist attacks in "In the Ruins of the Future" (37). As DeLillo elaborates, "The materials and methods we devise make it possible for us to claim our future. We don't have to depend on God or the prophets or other astonishments. We are the astonishment. The miracle is what we ourselves produce, the systems and networks that change the way we live and think" (37). Yet technology, too, blurs boundaries among humans, corporations, and machines and reshapes human conceptions of reality and time to serve capitalism. According to DeLillo, time, in the new millennium, "is scarcer now. There is a sense of compression, plans made hurriedly, time forced and distorted," a perspective that speaks to Judy Wajcman's argument in *Pressed for Time* ("In the Ruins" 39). And there exists a sense in DeLillo's writing that he sees technology as being part and parcel of what Manfred B. Steger, in his discussion of early views of globalization, refers to as a "techno-economic juggernaut" that spreads "the logic of capitalism and Western values" as though it is a "steamroller flattening local, national, and regional scales" (1). Instead of generating the kind of flat world that Friedman idealizes in *The World Is Flat* through his discussion of "old hierarchies" being "flattened and the playing field" being "leveled" predominantly because of emergent digital technology, digital devices as De-

Lillo sees them and global capitalism as digitization's bedfellow produce, to disconcerting ends, a flatness in life (44). This flatness functions as an intensification (to again reference Nealon's term from *Post-Postmodernism*) of Fredric Jameson's notion of "depthlessness," defined by Jameson as "a new kind of superficiality" that emerges as a result of the commodification of culture (9).

This analysis considers *Cosmopolis* as a philosophical meditation on the relationship between capitalism, digitization, and the physical and psychological experiences of being human in the cybercapitalist twenty-first century.[2] I argue that by setting the story on a day in April 2000, around the time of the dot-com bust, and by describing the personally and professionally suicidal actions of the twenty-eight-year-old computer-hacker-turned-currency-trading-billionaire Eric Michael Packer, DeLillo invites his readers to contemplate the hybridizing and flattening effects of digital devices: the ways in which they affect approaches to reading texts and seeing the world and people. He strategically texturizes, hybridizes, or flattens different concepts and characters in his fiction to draw attention to the ways in which digitization and capitalist forces that shape it have power to set the terms for material reality and psychological perspectives on that reality. In turn, he shows that works of fiction such as his novel retain power to rehabilitate human ways of thinking about the world. Ultimately, I suggest that DeLillo positions literary fiction as a countercapitalist, counterdigital, and political force that can equip readers to engage in meaningful modes of seeing and living in the world. Fiction allows them to think critically about what counts as reality, how they come to know information, and how they have come to inhabit the historical, global-capitalist moment in which they live. As a result, readers can envision a countercapitalist future in which humanity learns from missteps in history as opposed to adhering to false notions of American progress through technological development. And they can realize a future beyond the ruins that DeLillo references in "In the Ruins of the Future" by transcending the limits of commodification and the resolute grip of digitization.

In the second part of this chapter, I focus on Eggers, who has spent over two decades of his life near Silicon Valley and multinational tech-industry corporations such as Google and Facebook. Even though he relies on digital media to make a living as a professional writer and as publisher of the online *McSweeney's Internet Tendency* and the innovative print journal *Timothy McSweeney's Quarterly Concern*, he exhibits skepticism toward corporate-born digital media that resemble DeLillo's. He avoids maintaining Facebook or Twitter accounts, and he has no smartphone or home internet service. In a 2018 *Monocle* interview, he notes, "I've never been able to have Internet at home or cable TV or things like that because [...] I get distracted and I don't work much" ("Dave Eggers"). Instead, he goes to the library when he needs to conduct research. As he sees it, referencing the balance between natural and technological aspects of human life, "We all have to figure out how we feel most balanced," and we have to ask ourselves key questions about our relationship with digital technology, for instance, "are these tools that you are using or are you being used?" ("Dave Eggers"). Moreover, he suggests that Americans and citizens of the contemporary globe might benefit from assessing the degree to which they literally and metaphorically buy into the promises of digital technology. He sees digital and social media as dividing communities of which humans are a part as opposed to bolstering a genuinely connected, socially just world that he appears to desire by virtue of writing about human rights abuses in *What Is the What* and *Zeitoun* and sponsoring social justice work through Voice of Witness and 826 Valencia.[3] In Eggers's words to Paul Laity, "Social media separates and isolates us" and brings out "an inversion of every good quality" in individuals ("Dave Eggers: 'I always picture'"). These media help to create the alienation and disconnection that characterize the life experiences of Alan Clay, the tragic protagonist of *A Hologram for the King*, Eggers's first sustained meditation on digitization and economic globalization. In the novel, Eggers portrays Clay's failed attempts to market his American business's digital media products in a globalized world in the aftermath of the American Century. Through the novel, Eggers sets the stage for his subsequent and most developed meditation on digitization and globalization, *The Circle*, a work of dystopian fiction that tells the Orwellian story of twenty-four-year-old protagonist Mae Holland's developing relationship with the Circle, a Google-like American tech company. It employs Mae after her graduation from college; seduces its employees and users toward conformity through digital products; monopolizes the tech industry; and even supplants the American government as a false, profiteering purveyor of opportunities for democracy.

This analysis considers *The Circle* as a metaphorical and allegorical parable that draws attention to ways in which corporations reshape mass conceptions of individual identity, humanity, and reality through digital devices and media that they produce and sell. Emerging out of Eggers's observation that "*The Circle* is about abuse of power" and also out of Timothy W. Galow's argument that *The Circle* explores "the ways in which contemporary technology could potentially reshape the social order," I argue that Eggers portrays corporations and digital media much as DeLillo does: as sustaining a religious aura in that they function as powerful sources of near-fundamentalist American devotion (Eggers, "A Short Interview"; Galow 115). Tech companies such as Apple, which notably has circular headquarters in Silicon Valley, spread cybercapitalist fundamentalism as a digital-age extension of capitalist or market fundamentalism, defined by Malise Ruthven as a devotion to "global capitalism" that results in "deregulation and tight fiscal constraints on the economies of developing nations, with dire consequences for the poorest sections of society" (Ruthven 21). Much as capitalist fundamentalism fosters staunch dedication to capitalist ideals, cybercapitalist fundamentalism fosters devotion to the ethereal digital products that tech corporations produce and sell and to digital data, which demystify human life and the natural world. Ultimately, Eggers postulates that countering corporate and digital interests in what Friedman refers to as the flat world may exist as an impossibility if humans are unable to exhibit love for one another that rivals the love they sustain for corporate-born cyberspaces and the corporations with which they seek to merge. Much as DeLillo celebrates literature for its capacity to produce critical thinking, Eggers posits that print books and journals have power to counter the data-oriented and corporate mind-set that steers the United States toward a dystopian political reality. Through reading and understanding allegorical and metaphorical works of art and literature, they might counter toxic, inevitably superficial, and antiphilosophical urges to commodify knowledge with the simple goal of owning and knowing information. They might make manifest an alternate reality of radical individuals who form philosophical human communities that celebrate literature's mysteries and compassionate ways of understanding one another and the hybrid world.

Cybercapitalist Flatness and a Rehabilitated Future through Fiction in Don DeLillo's *Cosmopolis*

As DeLillo portrays it in *Cosmopolis* (2003), the focus of the first part of this chapter, American identity at the turn of the millennium exists as hybridized because it interconnects with capitalism and technology. On one hand, DeLillo shows interest in the relationship between the United States and capitalism, which work together to produce the contemporary globalized world. As Sven Beckert and Christine Desan suggest, "Just as understanding capitalism is essential to understanding the history of the United States, understanding the United States is essential to understanding the history of global capitalism" (3). According to Carl Degler, capitalist ideology "CAME IN THE FIRST SHIPS" to the New World (*Out of Our Past* 2). As a result of events such as the 1886 Supreme Court ruling in *Santa Clara v. Southern Pacific Railroad*, which accidentally declared corporations to be persons eligible for rights and protections by means of the Fourteenth Amendment to the United States Constitution, the relationship between Americans and capitalism continued to develop in a unique way. On the other hand, DeLillo shows interest in the relationship between the United States and digital technology. Like Pynchon, who showcases America's history with digital technology in *Bleeding Edge*, DeLillo gestures toward depictions of the United States as a technocracy throughout his oeuvre, for instance in *Underworld*, in which he employs DuPont's "BETTER THINGS FOR BETTER LIVING THROUGH CHEMISTRY" slogan as the title for the novel's fifth part (*Underworld* 499). The slogan speaks to what Carroll Pursell suggests is a World War II–era American notion that "technological progress, and the science that supported it, held the key to a stronger, richer, healthier, and happier America" (xii). Although the idea that technology makes American life better waxed and waned as the twentieth century rolled toward the twenty-first, literal and figurative investments in technology remained steadfast. They produced, to reference DeLillo's "In the Ruins of the Future," a "utopian glow of cyber-capital" (33). And they produced the technologically saturated nation that Eric Packer as DeLillo's protagonist both inhabits and helps create by working with a start-up company during the dot-com boom of the 1990s and then by thriving as the cybertrading CEO of Packer Capital until the boom turns into a bust in the year 2000.

DeLillo suggests that the fusion of American identity with technology and capitalism leads the Americans of *Cosmopolis*, especially the American elite, toward increasingly inhuman ends. For instance, Eric sees humans of the novel as zombie-like cogs serving the cybercapitalist system. They sit in their "cubicles exposed at street level, men and women watching screens" (75). More notably, however, as several critics of DeLillo's work have observed,[4] Eric himself functions as a posthuman or cyborg because of the degree to which he disconnects from the real and feels drawn toward digital and corporate phenomena. In his own words, he sees the "interaction between technology and capital" as the only "thing in the world worth pursuing professionally and intellectually" (23). He thus exists as a prime example of what Andy Clark theorizes as a "natural-born" cyborg (6). For instance, Eric merges with his digitally enhanced limousine and the technologically advanced digital watch he wears. His body comes to function as mechanized according to DeLillo's rhetoric, as evidenced, for instance, by the narrator's remark that Eric felt a street scene "enter every receptor and vault electrically to his brain" (65). In opting to journey across Manhattan along 47th Street in extraordinary traffic caused by the U.S. president's visit to New York in what David Cowart views as a travesty of "both Homeric hero and Leopold Bloom," Eric spends much of the novel absorbing reality through digital screens that line the interior of his limousine ("Anxieties of Obsolescence" 186). Eric's destination—Anthony Adubato's old-fashioned barbershop in the Hell's Kitchen neighborhood of his deceased father, Michael Packer, and of his own childhood—functions to underscore the peculiar, hybridized nature of his life. Although Eric sustains a single-minded focus on a physical experience of bodily care and although he engages in bodily pleasures by having sex with three different women along his ride, among them his estranged new wife, the poet Elise Shifrin, he simultaneously lacks physical concerns. He reveres virtual reality and finds mystery and existential meaning in "medleys of data on every screen, all the flowing symbols and alpine charts, the polychrome numbers pulsing" (13). He lives his corporate life torn between the pull of physical and virtual attractions and demands.

Throughout the novel, DeLillo considers ways in which cyborg citizens such as Eric read, process, and interpret digital information that purportedly functions to interconnect disparate people and nations even though it near-methodically renders them disconnected from one another. Hence he presents *Cosmopolis* as a philosophical and literary parable of literacy by underscoring Eric's problematic approaches to reading—approaches of the sort that N. Katherine Hayles explores in *How We Think: Digital Media and Contemporary Technogenesis*. Specifically, Eric engages in what Hayles refers to as hyper reading, which involves "skimming, scanning, fragmenting, and juxtaposing texts" and which functions as "a strategic response to an information-intensive environment" (12). In other words, Eric continually scans "[p]atterns, ratios, indexes, whole maps of information" and even sees them as transcendent—or as involving a near-metaphysical world: as "our sweetness and light," and as a "fuckall wonder" (DeLillo, *Cosmopolis* 14). Yet in reading digital data in this way instead of reading them more closely, Eric positions data as "pure spectacle," to cite the words of Vija Kinski, Eric's chief of theory (80). He positions them as a parody of the "sacred" in that they are "ritually unreadable," as evidenced by Eric's failed predictions of the Japanese yen's behavior (80). Although Eric sees an abundance of market data, he fails to read that data in a meaningful or productive way. He fails to go beyond the surface of "standard models" of information (21). And, perhaps most significantly, he fails to recognize that the problem is his way of reading and recklessly extrapolating on what he hyperreads.

For DeLillo, authors who philosophize technology and reading as he does have potential to create a "counter-narrative" of the sort that DeLillo mentions in "In the Ruins of the Future": a narrative such as *Cosmopolis* that invites readers to make meaning of the sort that eludes Eric ("In the Ruins" 34). Authors like DeLillo can invite readers to contemplate superficial ways of reading such as those that Eric engages in, and they can also present subversive ideas that fly under the metaphorical radar of a global capitalist system and hyperreaders. These authors exist as distinct from Elise as a fictionalized poet who sustains an "element of remoteness," "static," and a marriage to money of the kind that DeLillo scorns (*Cosmopolis* 16, 17). Instead, they resemble the Polish writer Zbigniew Herbert, whose "Report from a Besieged City," a critique of authoritarianism that Herbert encountered in Nazi and Soviet forms, perplexes Eric. Eric fixates on Herbert's image of a rat functioning as currency and fails to see that Herbert's anti-authoritarian ideas function as an unwelcome rat in his de facto authoritarian ideology. He fails to see that the kind of authoritarianism that Herbert bolsters through his association with cybercapitalist globalization renders him as the object of Herbert's critique. Along the same lines, these authors resemble the American visual artist Mark Rothko, who creates abstract expressionist paintings about quintessentially real human emotions—paintings that Eric appears to misread even though he sees himself as a committed Rothko connoisseur. Although Eric has an opportunity to own a Rothko painting and although he wants to buy the whole Rothko Chapel and put it in his Manhattan home, he never articulates that he has any sense of Rothko's Marxist motivations. He fails to recognize that Rothko believed that "[f]reed from a false sense of security and community, the artist can abandon his plastic bankbook, just as he has abandoned other forms of security" and create "transcendental experiences" (Rothko, "The Romantics" 58). And he fails to understand that Rothko located the Chapel in Houston because, as James E. B. Breslin puts it, he was "[t]roubled by the mid-1960s art market" and sought to withdraw from "the New York art world" (459). He wanted viewers to have to "make a kind of pilgrimage journey" away from art's economic center to see his spiritual work (Breslin 464).

In particular, DeLillo invites close readings of his own novel as a counternarrative to unearth problems of globalized reality instead of simplifying globalization and allowing its advocates to level two divergent narratives of it into an unequivocally positive and thereby flattened narrative. Whereas proponents of globalization laud it for interconnecting apparently disconnected people and places across the globe and see it as producing, for instance, the type of success story and hybrid identity that Raymond Gathers of the Bronx embodies as the self-made Sufi rapper Brutha Fez in *Cosmopolis*,[5] critics of the phenomenon including DeLillo censure it for what Steger calls "economic deregulation and a culture of consumerism on the entire world" (Steger 135). Over the course of the novel, DeLillo repeatedly draws attention to evidence of the deeply harsh circumstances that victims of the globalized flat world encounter. As Martina Sciolino suggests, DeLillo juxtaposes "rows of nearly indistinguishable white limousines" with diverse drivers who speak an array of languages and come from apparently less privileged nations that resemble Saeed and Nadia's home in Mohsin Hamid's *Exit West* (Sciolino 225). And as Jung-Suk Hwang explains, DeLillo, too, draws attention to bodily evidence of suffering among minor characters of color in the novel to show that "the utopian vision of cybercapitalism continues to produce the poor Other" (38). For instance, through Ibrahim Hamadou, Eric's limo driver, who has a "collapsed eye," and through Elise's Sikh taxi driver, who is "missing a finger," DeLillo exposes globalization's often obscured backlash, which Zadie Smith also explores through her portrayal of the Felixstowe poor in "Meet the President!" (DeLillo, *Cosmopolis* 163, 17). Although

DeLillo never reveals the apparently trying experiences that shape these characters' lives and bodies, he leads readers to imagine the existence of a connection between Eric's privilege and their privations because all privilege in the interconnected, globalized world exists in symbiosis with suffering. As Sciolino points out, DeLillo invites readers to treat Hamadou's scarred face as "a text" that is open to interpretation (228). And he bids readers to see through the experience of reading what Sciolino calls "the neoliberal order of things"—an experience that ideally prompts readers to resist cultural illiteracy and the impulse toward detached apathy that flat screens and digital devices perpetuate (223).

In turn, DeLillo flattens or blurs distinctions between apparently disparate elements of existence in his text to underscore the fact that superficial approaches to seeing and reading as digital technology encourages them have material effects that serve global capitalism alone. For instance, he portrays professional and personal endeavors as indistinguishable to showcase the way in which the disappearance of a work-life balance for Americans serves corporations, not people. Whether Eric spends his day working or playing out of his limousine remains up for debate, a point that is spotlighted by Richard Sheets, aka Benno Levin, Eric's disgruntled former employee at Packer Capital and his antagonist. Benno observes that Eric "voided" his office "to work elsewhere, or work wherever he happened to be, or work at home in the annex because he did not really separate live [sic] and work, or to travel and think, or to spend time reading in his rumored lake house in the mountains" (56). In the limousine, Eric works in a quite different way than Paul O'Rourke does by the conclusion of Ferris's *To Rise Again at a Decent Hour* (when Paul works with the humanitarian goal of realizing the social good). He works for his own personal good, meeting with a range of advisers including Shiner; Michael Chin, his currency analyst; Jane Melman, his chief of finance; and Kinski. In large part, Eric discusses work-related matters with his staff. Everyone talks with him about his reckless bets against the Japanese yen, and everyone except for Kinski urges him to stop betting against it. Yet an element of play also exists in his efforts to work, perhaps because of the privilege he has that working-class workers lack. He continually interrupts professional exchanges with personal ventures, for instance sex inside his limousine and along his limousine's route, lunch with Elise, a medical exam inside his limousine, and a late-night stint as an extra in a movie shoot. Evidence of Eric's investment in play, too, exists in his approach to bankrupting Packer Capital and Elise's European family's banking fortune. By merely clicking buttons on his watch, phone, and computer as a self-absorbed child at play with a video game clicks them, Eric regains some semblance of the childhood he lost when his father died of cancer when he was only four years old. He sees digital devices that exhibit data as "playthings" (171). As the narrator describes one instance of his hasty, selfish, and childlike currency trading, "He sat down long enough to take a web phone out of a slot and execute an order for more yen. He borrowed yen in dumbfounding amounts. He wanted all the yen there was"—even if his actions wreak havoc on the global market (97).

Along the same lines, DeLillo suggests that distinctions between elite characters of different kinds flatten or dissolve in the digital age to ends that serve corporations which seek to influence governments, if not outright function as they do. In particular, DeLillo creates confusion between Eric, whose name appears within the word Am*eric*a, and the American president Midwood. Much as Andy Warhol flattens out distinctions among different kinds of fame by creating aesthetically similar artistic renderings of, for instance, the popular musician Elvis and the totalitarian politician Mao Zedong, DeLillo suggests that noteworthy similarities exist between Eric and Midwood and, in turn, between corporations and American government. For instance, he showcases Eric and Midwood as objects of digital-age American veneration by virtue of the fact that both have livestreamed their lives for mass consumption over the internet. As the narrator observes when Eric is distracted by an image of Midwood on a screen in his limousine, "the chief executive on live videostream, accessible worldwide," is a "feature of the Midwood administration" (76). Similarly, readers learn that Eric livestreamed his life in the surreal space of the World Wide Web. As Benno describes it, "I watched the live video feed from his website all the time. I watched for hours and realistically days. What he said to people, how he turned so sharply in his chair" (151). Markedly, DeLillo flattens out distinctions between Eric and Midwood by portraying both executives as contending with credible threats against their lives. Whereas DeLillo introduces the credible threat against Midwood early in the novel through a conversation between Eric and Torval, Eric's bodyguard, he introduces the credible threat against Eric later in the novel—in the first of two chapters written from Benno's perspective. As Benno describes making the threat against Eric, "I made a phone threat that I didn't believe. They took the threat to be credible, which I knew they had to do, considering my knowledge of the firm and the personnel" (56).

Moreover, DeLillo represents similarities between corporate elites and lower-class citizens who oppose them to showcase ways in which global capitalism realizes Marxist theory in paradoxical ways by rousing its own nominal and more formidable opposition: gravediggers that Kinski and Eric reference in their conversation about Karl Marx and Friedrich Engels's "Manifesto of the Communist Party." According to the manifesto, the "development of Modern Industry [. . .] cuts from under its feet the very foundation on which the bourgeoisie produces and appropriates products" because the bourgeoisie produces "its own grave-diggers," thereby making its "fall and the victory of the proletariat" inevitable (Marx and Engels 483). Throughout the novel, DeLillo presents different prospective gravediggers, for instance the Times Square protesters, who, at least according to Kinski, "invigorate and perpetuate the system" that they claim to oppose in that they exist as what Kinski sees as "an appropriation" or simulation of apparently real protesters of old, for instance Thích Quảng Đức, the Vietnamese Buddhist monk who set himself on fire in 1963 (*Cosmopolis* 90, 100). Indeed, these contemporary protesters may hope and yet fail to function as gravediggers because they lack an understanding of the nuances of Marxism as it purportedly shapes their position, as demonstrated by their misappropriation of the first line of the communist manifesto. In hacking an electronic display to write "A SPECTER IS HAUNTING THE WORLD—THE SPECTER OF CAPITALISM," they showcase their failure to see that Marx and Engels intended to galvanize a heretofore unorganized resistance to capitalism by way of writing that a "spectre is haunting Europe—the spectre of Communism" (DeLillo, *Cosmopolis* 96; Marx and Engels 473). In other words, these protesters misread Marx and Engels much as Eric misreads Herbert's poetry, which the protesters likewise read, further reinforcing their similarity to Eric.

By contrast, DeLillo showcases similarities amid superficial differences between Benno and Eric. He characterizes Benno as an impoverished and unemployed former assistant professor of computer applications at a community college and former currency analyst who scorns technology and lives "offline now" (149). Benno reveres material features of physical life much as the protagonist of Shteyngart's *Super Sad True Love Story* does, but to a more intense if not disturbing degree, as illustrated, for example, by the fact that he "used to lick coins as a child" (*Cosmopolis* 154). He mentally lives in the past, dwelling on the loss of his wife and even lacking a sense of his own age.[6] And he squats in an abandoned building on the west end of 47th Street in the Hell's Kitchen neighborhood where Eric began his life and grew up—a neighborhood that Aristi Trendel positions as a parody of the "American Frontier" because it lacks promise that the American West afforded (124). By contrast, DeLillo characterizes Eric as a wealthy currency trader who worships technology and indulges in the pleasures of virtual as opposed to material reality, as evidenced by his disdain for the "grain" of the 47th Street Hasidic diamond district, a manifestation of old American capitalism, which he views as "an offense to the truth of the future" (DeLillo, *Cosmopolis* 64, 65). Eric lives in the future much as Jennifer Egan's Lulu Peale does in *A Visit from the Goon Squad*. He scorns "aged and burdened" technological devices such as ATMs, which Benno loves, and resides in a quasifuturistic space: a tower on the east end of 47th Street that resembles bank towers, which Eric sees as being "in the future, a time beyond geography and touchable money and the people who stack and count it" (54, 36). Yet by way of limiting identifying markers in dialogue between the two men in the novel's fourth chapter, DeLillo renders them as relatively indistinguishable doppelgängers—twins evocative of those described in the twin paradox that physicists reference in discussing Albert Einstein's Special Theory of Relativity,[7] a theory of time that appears to enthrall DeLillo (as evidenced by his recent attention to it in *The Silence*) and that Eric reads in the novel's opening pages. Eric and Benno each has an asymmetrical prostate. Each has unraveled and entered into poverty if not madness as the result of a failure to project the behavior of currency.[8] And each perhaps functions as a gravedigger of the kind that Marx and Engels describe. Eric as a capitalist gives birth to Benno, metaphorically speaking, by firing him, and Benno attempts to bury capitalism metaphorically by virtue of threatening Eric's life. Similarly, Benno gives metaphorical birth to Eric by threatening his life in a way that fuels Eric's recklessness. As the narrator puts it, "the credible threat was the thing that moved and quickened him"—even though his invigoration moves him, paradoxically, to devastating ends, as demonstrated by Eric's attempts to bury capitalism by depleting his fortune and wreaking havoc on the global market in the process (107).

The peculiar connections, flattened-out distinctions, and hybrid realities that appear throughout *Cosmopolis* give the novel a surreal quality that speaks to the surreal real-world experiences that DeLillo thinks Americans may be more prone to have in the digital age. Certainly, the virtual realities that digital devices allow contemporary citizens to inhabit have dreamlike or even transcendental qualities in contrast to material reality, as illustrated, for instance, by Eric's experience of looking at data on a screen early in the novel. As the narrator describes it, Eric sees data as "soulful and glowing" (24). They exist as a means by which he can transcend mundane aspects of material reality. Digital devices, too, have power to render reality as surreal if not outright dreamlike because of the ways in which they inhibit sleep for natural-born cyborgs who merge with them. As Marie-Christine Leps suggests, Eric is "[l]ike cybercapitalism" that runs 24/7 because, as cybercapitalism's embodiment, he "cannot sleep" (309). As numerous studies suggest,[9] the overuse of digital devices leads to sleep disturbances, which DeLillo indicates Eric has in the opening lines of his novel when he notes that sleep eluded Eric "more often now, not once or twice a week but four times, five" (*Cosmopolis* 5). For Eric, night develops into a "running lull," and as DeLillo suggests, day resembles night because of Eric's exhaustion and resultant psychological oblivion (25). Not only do details such as "the Latin name of the tree" he sees outside of Didi's apartment elude him, but he loses articles of clothing including his undershirt and jacket as he skips between near-ethereal intimate encounters that resemble sex dreams (25). Certainly, the whole of DeLillo's novel might be conceptualized as an extended dream sequence that complements contemporary reality in part because of the otherworldly nature of, for instance, Elise, who seems to near-magically appear repeatedly along Eric's crosstown odyssey. It also might be conceptualized as an extended dream sequence because of Eric's sleep-deprived state, which DeLillo underscores by portraying Eric as finally arriving at his barbershop destination in the dead of night and, while getting his hair cut, beginning to "fade, to drop away" as a child might (165).

In rendering the novel as akin to a dream sequence, a technique typically employed in movies, DeLillo intimates that peculiar and perhaps unsettling connections exist between movies as the originators of screen culture in the United States and future-oriented, technologically savvy Americans in the digital age. As Eric reveals to Benno in the novel's concluding sequence, movies come to supplant his dead father as a presence in his life. They serve as a means by which he and his mother can address or perhaps avoid addressing their grief or their changed relationship and be "alone together," to reference both Eric's own language and the title of Sherry Turkle's third book on the social effects of computers (*Cosmopolis* 185). They serve as the forebears of contemporary digital screens with which Eric sustains an apparent love affair. Perhaps because of Eric's intimate history with movies, he feels drawn to an outdoor movie shoot, which spotlights the long history that screens have of flattening humanity, human experience, and meaningful human connections while creating the illusion of connectivity. Although "financing has collapsed" for the film (alongside and perhaps because of the collapse of Eric's fortune, which may well have been funding the film), Eric feels a sense of connection in removing his clothing and joining hundreds of naked extras in the city street to appear in a master crane shot and a handheld digital video shot that will likely never make it into a movie theater (175). Paradoxically, the prospect of the big screen, where godlike celebrities glitter, affords him the chance to descend out of his elite social class to be among ordinary people. As the narrator suggests, he wants to be "among them, all-body, the tattooed, the hairy-assed, those who stank" and in ways "one of them" (176). And he even magically encounters Elise in the crowd of extras and takes the opportunity to belatedly and futilely consummate his marriage in a scene that the narrator notably describes as resembling "a black-and-white film that was being screened in theaters worldwide"—an unreal scene out of history that DeLillo may characterize as such because Eric's marriage has no future (177). Much as the movie funding collapses and renders Eric's apparently meaningful moment of connection meaningless in the grand scheme of things, so, too, does Eric's marriage collapse in a revelation of meaninglessness for readers if not for the deluded and utterly devastated Eric. It apparently ends as Elise walks "head-high, with technical precision" of the sort that Eric reveres in digital devices (178). She walks out of Eric's life forever.

Moreover, by way of concluding events in the novel, which culminate in a standoff between Eric and Benno that is evocative of the one between Jack and Willie Mink at the end of *White Noise*, DeLillo suggests that the line dividing fiction and reality (or the reality presented within the confines of a fictional movie from the reality represented within the confines of a literary fiction novel) is blurry because movies influence perceptions in the globalized and digitized times. Perhaps because of his devotion to screen culture in its different forms, Eric behaves increasingly more as a screen-based, fictional movie character might as the novel progresses toward its enigmatic conclusion. He behaves as though his life is a circumscribed movie on a screen designed for the primary purpose of entertainment. First, he shoots Torval with Torval's own gun without concern for the consequences of murder, and in doing so, he moves the plot of DeLillo's novel forward, making it resemble the plot of a blockbuster murder movie. For his act of violence, he might even become the subject of television news programming, which airs on screens in his limousine and glorifies and numbs the masses to violence. Later, while reflecting on how, against all odds, "a single kick suffices" to open doors in movies he watches with his mother after his father's death, Eric further conflates reality and movies when he kicks in the door to Benno's home and watches the way it "opened at once" like kicked doors open in movies (184, 186). Eric's comments and actions underscore key revelations toward which DeLillo aims to move his readers: first, that the novel as a work of fiction has a long history of coming into conversation with screen culture despite widespread late twentieth- and early twenty-first-century claims that digitization threatens books in a new way,[10] and second, that fiction in its various forms always already exists in dynamic interplay with reality. As Benno explains his own conceptualization of the relationship between reality and fiction, "Whether I imagine a thing or not, it's real to me," a point that writers of fiction such as DeLillo would likely appreciate (192). "The things I imagine," Benno continues, "become facts" (192). As a result, although Eric sees the crime that Benno apparently wants to commit as a "cheap imitation" and "stale fantasy," Benno indicates that he sees his plot as wholly part of reality by saying that his plot is "all history" (193).

DeLillo's readers never come to a clear sense of what counts as reality in *Cosmopolis* per se because, for DeLillo, reality is relative and diverges from laws of physics, and because mystery exists as DeLillo's rhetorical and countercorporate goal.[11] Most notably, readers come to question the reality of Benno's innocence as well as his guilt when they encounter his mysterious declaration of Eric's death in the "Night" section, which comes chronologically before the "Morning" section and therefore contributes to the countercorporate mystification of time in DeLillo's novel. Much as it does in *Underworld*,[12] time cycles simultaneously in two directions in *Cosmopolis*: it runs forward in the longer sections told from Eric's perspective, and it runs backward in the two short sections told from Benno's. As a result, DeLillo comes into conversation with Einstein's Special Theory of Relativity by generating grounds for a "new theory of time" that Kinski says humanity needs—perhaps because, to reference words spoken by Eric that echo DeLillo's words in "In the Ruins of the Future," "[t]ime is a thing that grows scarcer every day" (*Cosmopolis* 86, 69). Whereas Einstein suggests that time is bound to space even though individuals in space might experience events at different times and may experience the effects of time in different ways, DeLillo intimates that time emerges as mysteriously disjointed, perhaps as a result of the fast, productive, and corporate-serving pace of life that Wajcman explores in *Pressed for Time*. As Benno puts it in this section's first line, Eric is "dead, word for word" even though he lives his fast-paced life in his absurdly slow-moving limousine in the sections that precede and follow the remark (55). According to the Special Theory of Relativity, the faster one travels, the slower one ages and, presumably, the later one dies. But the peculiar blend of paces renders Eric as apparently dead. And he is apparently dead even though readers never see the moment in real time and space at which he dies and even though readers lack irrefutable evidence that Benno is his killer.

DeLillo, too, mystifies reality within the world of his fiction by virtue of the lexical choices he makes and questions regarding reliability, both of which function to create feelings in readers similar to those that Pynchon creates through his representation of a possible conspiracy in *Bleeding Edge*. Benno's words about Eric's death as existing in words draw attention to the quintessentially textual as opposed to physical nature of Eric's death in the incomplete and unreliable text that Benno produces: a confession by genre that spans roughly fourteen pages as opposed to the ten thousand pages that Benno initially claims he wants to write. Disparity exists between Benno's perception of himself and the reality of his authorial abilities. It exists between what he aspires to write and what he is actually capable of writing. Throughout his narrative, Benno spotlights his own instability and unreliability as a narrator, for instance indicating that he suffers from two illnesses that may well inhibit his view of reality. As Benno explains, he suffers from "*hwabyung*," or "cultural panic," and "*susto*," or "soul loss," both of which he contracts from the internet, a literal source of unreliable information that may infect Benno with unreliability (56, 56, 152, 152). More notably, Benno outright reveals himself to be a liar. In his words, "I used to tell the truth. But it's hard not to lie. I lie to people because this is my language" (150). Benno even acknowledges that the threat he makes against Eric is "mostly empty," but because Benno is a liar, this statement remains just as questionable as Benno's testimony of Eric's death and everything and anything he says in the novel (58). As a result, DeLillo's readers, much as the readers of Pynchon's *Bleeding Edge*, have no sure footing. Benno may be mad, plain and simple, or he may be a killer. Whether or not Benno manages to "rise up from the words on the page and do something, hurt someone" is never entirely clear, especially given that he never appears to achieve a "depth of writing" that he claims will tell him whether he is "capable" of real violence (150).

Readers of DeLillo's work thereby realize that DeLillo is most interested in the truth of uncertainty much as Ferris is interested in it in *To Rise Again at a Decent Hour*. He is interested in what runs counter to near-ubiquitous digital information in an age in which digital citizens readily relinquish all doubt and see everything that Google reveals as true. For DeLillo, the uncertainty that writers of fiction create by way of presenting textured and complex ideas in their texts may serve them well as they attempt to make sense of an increasingly hybridized reality. Fiction helps twenty-first-century humanity to read and navigate uncertain information, especially of the digitally mediated variety, because reality is shaped by subjectivities that fiction explores and not just by digital screens that serve corporate interests. In Benno's words, it "is what people think they see in another person that makes his reality" (57). Along the same lines, for better or worse, digital devices, as Kinski observes, are now "crucial to civilization" because they help "us make our fate" (95). They exist as new facts of life and inform ever-changing and thus uncertain human subjectivities. For instance, as Eric watches himself on a screen projecting a spycam-captured image in his limousine that he knows operates in "real time," or is "supposed to," he feels "his body catching up to the independent image" (52). More notably, when Eric finds himself amid an anticapitalist protest in Times Square, a corporate space that through its name makes reference to time, he feels the peculiar effects of digital-age, corporate, fast-paced time. He sees the effects of a bomb detonating on-screen before he feels them in real life. As the narrator describes it, Eric "saw himself recoil in shock. More time passed. He felt suspended, waiting" (93).

For Eric, who purportedly lives in the future, and for digital citizens who resemble him, real life feels as though it consistently lags behind and even imitates what appears in digitized, screen-based life—even if digital devices lack the capacity to represent truths that corporations want them to be able to represent for the sake of making a profit. Notably, a digital device, Eric's wristwatch, provides the only evidence in the Eric-centered narrative of Eric's perhaps merely virtual death. It provides evidence that DeLillo certainly sees as problematic because the truth that the watch presents resembles Benno's questionable truth or the truth of a cunning fortune-teller's crystal ball. The screen in Eric's wristwatch shows Eric his supposed future via an embedded electron camera that displays images on a crystal. It transcends its analog antecedent by providing more than a mere analogy for the movement of time. Too, it allows its user to transcend conventional limits for the work of attentive seeing or watching—toward which the word *watch* speaks—by showing Eric his own as yet unrealized death. According to Randy Boyagoda, the narrator of *Cosmopolis* echoes the end of *Underworld*, in which Sister Edgar experiences a "digital afterlife" through cyberspace (Boyagoda 23). As the narrator of *Cosmopolis* remarks, Eric glances "at his watch" and sees "a face on the crystal" that morphs into a body on the floor and then "the inside of an ambulance, with drip-feed devices," and then "a series of vaults" (204, 204, 206, 206). As the narrator continues, when one of the vaults—evidently in a morgue—slides open, Eric sees "the tag in tight close-up" and knows "what this meant" even though he acknowledges that he "didn't know how he knew this. How do we know anything?" (206).

The question that Eric asks about knowledge and certainty as he faces the prospect of his own death ultimately dovetails with the question that DeLillo wants his readers to contemplate through this metafictional moment about spiritual death at the hands of digitization: a question about the degree to which corporate-born digital screens have set the terms for what counts as reality. DeLillo not only invites readers to contemplate how they know Eric is dead but how they know anything at all in an era in which screens continually mediate their relationship with the real. Although the narrator observes that Eric wants to "trust the power of predetermined events" such as those that his watch appears to foretell much as readers may want to trust indications of Eric's death in the text, the purported future Eric sees may come at his own hands (147). He shoots himself in the hand and may well kill himself as well. Or the future Eric sees might not be his future at all. What he sees may be a hallucination because he is a delusional and sleep-deprived megalomaniac who is experiencing the traumatic repercussions of having committed murder. In reality, the future is not set, and Eric and DeLillo's readers alike retain agency to shape the realities they inhabit—even if they opt against exercising that agency. Thus, if Eric dies by novel's end, he dies in part because he fails to own his own life by living it. He dies because he already sees and perhaps for many years has seen himself as divorced from the living and the real. To cite Benno's words, he is "already dead" (203). Or rather, according to Benno, he behaves "like someone already dead. Like someone dead a hundred years. Many centuries dead" (203). Moreover, if Eric dies by novel's end, he dies in part because he believes that a screen-based existence in and of itself constitutes real life and perhaps even allows for the everlasting life that Kinski evokes when she describes digital disks as alternatives to tombs.[13] He lacks what the novel's narrator calls "the sheer and reeling need to be," and, I would add, to be or exist in the physical world (209). As DeLillo presents it, therefore, Eric serves as a hyperbolic metaphor for real-life digital citizens around the turn of the millennium who live life as barely alive because they conflate webbed, screen-based life with real life in metaphorically life-threatening ways. He serves as a metaphor of the problem of living in a liminal state between the web and the world, between life and death, as a result of overreliance on digital devices and media.

DeLillo uses the fiction of Eric's death that appears within his fictional novel to suggest that individuals who are owned by their digital devices are susceptible to being owned by history—even if they appear to revere the future and hate history on the surface. Notably, Eric exists as trapped by history because he is incapable of seeing history's potential utility. Although Eric marries Elise because he feels that "a little history is nice," he never finds a means by which to view history as interesting or useful, as evidenced by his remark that Kinski would "disappear" if she were to have "a history" (120, 105, 105). He never sees history as a foundation on which to build a trajectory toward a meaningful and reflective future. Instead, he squanders any potential for a meaningful relationship that he might have in his marriage much as he squanders his fortune. And he seeks out experiences from his own history without nostalgia (a phenomenon that Pynchon and Egan problematize in their respective novels on digitization),[14] but with the goal of fruitless and notably mechanistic repetition that is evocative of the repetition of ones and zeroes that form the foundation for all digital technology. For instance, in Adubato's barber shop, a space that provides a trip down memory lane for Eric, Eric indulges in knowing the future because the future exists as a mere re-creation of historical experiences with which he is familiar, as illustrated by the narrator's remark that "Eric knew what" Adubato "would say when he opened the door," or by the narrator's assertion that Eric had heard Adubato tell the same story about his father's cancer "a number of times" using "the same words nearly every time" (160, 160, 161, 161). As the narrator continues, this is what Eric "wanted from Anthony" (161). And what he wants from the visit to the barbershop more than he wants the haircut that he never even fully acquires is to feel "elapsed time" as it "hangs in the air"—even though he recognizes that the corporate power he embodies and seeks "works best when there is no memory attached" (166, 184). Eric thereby exists as a paradox when it comes to his views of history and the future. Paradoxically, fate as the watch represents it to Eric becomes a part of Eric's history. So when Eric engages with the watch's representation, he does so much as he engages with Adubato: with an interest in repetition alone, even if it involves his own death. He fails to see what the novel's narrator and reader see: that the end of the novel "is not the end" because Eric is "dead inside the crystal of his watch but still alive in original space, waiting for the shot to sound" (209). He fails to see that he can arguably make manifest a different future by acting on the "enormous remorseful awareness" of his mistreatment of Benno and "others down through the years, hazy and nameless" (196).

DeLillo implies that readers who free themselves from the circumscribed reality that screens present have the power to learn from historical narratives or fiction about history as a concept in order to forge a new, countercorporate, and rehabilitated future and a rehabilitated model of citizenship that counters what Jerry A. Varsava calls Eric's "false model" (105). To paraphrase and appropriate Kinski's words, capitalists support the destruction of the past in their efforts to construct the future.[15] They repeat mistakes of old much as Eric repeats history because they avoid valuing history for the ways in which it reflects the depth of human experience. They fail to see that history and fictional works that grapple with it as a concept have instructive potential. By contrast, readers of fiction who seek to counter the capitalist system that DeLillo critiques can do so much as writers of fiction such as DeLillo do via their negotiations with history—negotiations such as those that DeLillo portrays in "The Power of History," a 1997 *New York Times* essay that describes DeLillo's research into the 3 October 1951 National League pennant playoff game between the New York Giants and the Brooklyn Dodgers, the subject of the prologue to *Underworld*. Although DeLillo acknowledges in the essay that fiction "*is all about reliving things. It is our second chance*," he appears to see fiction and life alike as involving more than cyclical repetitions or re-creations of events ("The Power of History"). He explains that a contemplation of history exists as an opportunity for rethinking a historical moment or phenomenon. In his description of his own process, he says: "I found myself thinking about the event in a different way, broadly, in history, as an example of some unrepeatable social phenomenon, and I couldn't shake the impact of the game's great finish—the burst of jubilation in the old Polo Grounds and throughout much of the city when Bobby Thomson of the Giants hit the game-winning home run" ("The Power of History"). Furthermore, as DeLillo observes, "*The past is great and deep. It can make a writer expansive, open him to perspectives and emotions that his own narrower environment has failed to elicit*" ("The Power of History").

Hence, to borrow DeLillo's words from "The Power of History," when writers of fiction write about past events or about history as a concept, literary language emerges as "*a form of counterhistory*" that can "*shape the world*" and "*break the faith of conventional re-creation*" ("The Power of History"). Literary language emerges as a means by which to involve imagination in conceptualizations of past and future time. It prompts readers to engage in rehabilitative anti-establishment ways of thinking, and, specifically, it provides them with an alternative to thinking "past what is new," to cite Benno's gloss of Eric's unsettling and wholly corporate approach to time (*Cosmopolis* 152). Moreover, it allows them to contemplate in their own lives the question that Benno asks of Eric when Eric willingly walks into the home of his prospective assassin and a mystery that resides at the heart of De-

Lillo's novel: "Why are you here?" (190). By contemplating this metafictional question, DeLillo's readers might see the devastating end that awaits them if they allow corporate forces to own, authorize, and disseminate information and also flatten life without critique in a digital-media saturated and ever globalizing historical moment. More importantly, they might attain a sense of ownership over their own textured place in history and their lives so they can see themselves as more than mere commodities in a capitalist system. Indeed, by way of rethinking, as DeLillo rethinks them, time, reality, capitalism, and digitization as well as the potentially toxic relationship that they have with one another and with humanity, DeLillo's readers have an opportunity to counter the desperate feeling that Benno has: the feeling that, as Benno puts it, "My life was not mine anymore," and, more problematically, that Benno "didn't want it to be" (153). As a result, digital citizens who develop a more literary imagination through the practice of reading fiction can learn from missteps in history so that humanity's narrative can avoid ending "in the rubble," to reference DeLillo's poignant remark in "In the Ruins of the Future" (34). They can make genuine progress toward a perhaps inevitably globalized future that counters flatness by resonating with an enduring depth of human experience and feeling.

Cybercapitalist Fundamentalism, Digital Data, and Philosophical Understanding in Dave Eggers's *The Circle*

In *The Circle* (2013), the focus of the second part of this chapter, Dave Eggers portrays the Circle as a hybrid space that resembles the limousine of DeLillo's *Cosmopolis* for its luxurious and transcendent features. He describes it as resortlike and as a parody of a religious institution. Through his descriptions of its idyllic spaces, he purposefully blurs the line that divides corporate and religious culture to join DeLillo in prompting Americans to reflect on ways in which they deify capitalism. When Mae arrives at the vast campus as a new entry-level employee on a day evocative of 11 September 2001 for its "spotless and blue" sky—a day known for fanatical terrorist violence that showcased al Qaeda's hostility toward America's role in producing global capitalism—she observes that the "soft green hills," Calatrava fountain, picnic area, tennis and volleyball courts, childcare center, and "workplace, too," are "heaven" (*The Circle* 1). In part, her conspicuously religious rhetoric about the campus's material evidence of capitalist excess emerges because of her lower-class Fresno roots and the burden of student loan debt she carries from obtaining her undergraduate degree from Carleton College, a pricey private institution in Minnesota. And the religiously charged term she opts to employ foreshadows the metaphorical relationship between capitalism and religion that Eggers establishes through subsequent descriptions of the Circle as a quasimonastic space where dorms allow employees to live where they worship capitalism through their work. According to Eggers, the front hall of the corporate headquarters is "as tall as a cathedral" where worshipers gather (3). Eggers notes that religious artwork adorns the surfaces of spaces associated with historical periods. In giving a tour of the space, Annie Allerton, Mae's former college roommate and upper-level Circle employee, directs Mae's attention to a "stained glass" rendering of "countless angels arranged in rings" on the ceiling of what appears to be a Byzantine-era-themed room (27). Much as the screens and barriers of the Latin Mass mystify aspects of DeLillo's oeuvre for Amy Hungerford,[16] digital screens pepper the Circle and define it as a paradoxical space that involves both mystery and a purportedly "clear and open" character, particularly via "glass rooms" and surfaces (Eggers, *The Circle* 242, 29).

Furthermore, Eggers suggests that corporations in the United States and in the globalized world function rhetorically much as religious communities function according to Frank J. Lechner and John Boli, who observe that religion long predates "the current phase of globalization" with its efforts to globalize by spreading universal messages to "new adherents" (387). Eggers positions the Circle's three founders, Eamon Bailey, Tom Stenton, and Ty Gospodinov, known as the Three Wise Men, as analogous to the Magi in the story of Christ's birth. Too, he positions them as parodies of religious ministers spreading the Circle as a religion at best, or, as Mae's technophobic ex-boyfriend Mercer Medeiros suggests, as a "cult taking over the world" at worst (*The Circle* 260). They subsume Google, Facebook, and Twitter, converting these companies' innovations into profits for themselves (23). And they offer slogans as "core beliefs" to employees who function as new religious adherents in order to create cohesion in workers' sense of the company's mission, which Bailey often relates to social justice as though he is a religious missionary. For instance, a Circle employee named Dan illuminates for Mae that Circle workers believe in "*Community First*" (47). In turn, the Circle offers customers engagement with what Eggers positions as a satirical version of a universal message: Ty's groundbreaking invention, TruYou, a "Unified Operating System," at least according to Eggers's apparent misunderstanding of what an actual operating system is and does (21). The Circle offers them a technological platform that is equivalent to the one true God: a platform that combines "everything online that had heretofore been separate and sloppy" including "users' social media profiles, their payment systems, their various passwords, their email accounts, user names, preferences, every last tool and manifestation of their interests" (21). The Circle thereby offers them "simplicity, efficiency, a clean and streamlined experience" as well as a more "civil" online environment that reflects the kind of better world that religious believers strive to create (22).

In particular, Eggers suggests that tech companies such as the Circle spread the good news of digital devices and their data through capitalist means much as evangelical fundamentalist Christian ministers might spread the good news of Christ. In other words, Eggers complements Ferris's and Foer's respective representations of religion's potential utility in the digital age by suggesting that religious impulses emerge as toxic when the Circle works to convert everyday digital citizens into cybercapitalist fundamentalists who sustain a single-minded desire to attain information. Eggers builds on definitions of religious and capitalist fundamentalism to develop a fictionalized cybercapitalist fundamentalism: a philosophy of the merger of digital technology and capitalism that speaks to existing philosophies of technology. Whereas scholars of religion characterize religious fundamentalists as demonstrating staunch faith and a tendency toward literalist readings of religious texts (readings that refuse to recognize the possibility that religious texts' authors may have employed literary devices such as metaphors, allegories, or symbols), Eggers characterizes cybercapitalist fundamentalism as involving staunch faith in digital devices that are produced by capitalist fundamentalists, staunch believers who, like Eric Packer, believe deeply in global capitalism as an economic system. Eggers, too, characterizes cybercapitalist fundamentalism as involving a deep desire for and unwavering faith in information that digital devices share with users—faith that the narrator points toward in describing Mae's revelation of her desire for information. In the narrator's words, "It occurred to her, in a moment of sudden clarity, that what had always caused her anxiety, or stress, or worry, was not any one force, nothing independent and external—it wasn't danger to herself or the constant calamity of other people and their problems. It was internal: it was subjective: it was *not knowing*" (195). Certainly, Bailey as a "Midwestern church-goer" turned metaphorical corporate preacher functions as an archetypal cybercapitalist fundamentalist (293). He believes deeply in "the perfectibility of human beings" and the notion that demystification of human behavior until there are "no secrets" in life will give humans "no choice *but* to be good" (291, 292).

Eggers suggests that cybercapitalist fundamentalists venerate information as it is obtained through digital surveillance and other means and as digital devices present it because information promises digital citizens religious experiences that may otherwise elude them. For instance, through engagement with digital products that the Circle relies upon and markets, the Circle's employees and customers are promised a version of everlasting life that resembles the life after death that Sam Bloch gives to Isaac Bloch, his dead grandfather, in Foer's *Here I Am*. As a Circle employee says to Mae upon transferring data from her personal digital devices to corporate devices that she is urged to use in her new position, "Now everything you had on your other phone and on your hard drive is accessible here on the tablet and your new phone, but it's also backed up in the cloud and on our servers. Your music, your photos, your messages, your data. It can never be lost" (43). Similarly, social media participation offers Circle employees and customers a parody of transcendence. In response to critiques her colleagues Josiah and Denise make of her lack of engagement with social media, which the Circle characterizes as part and parcel of work as opposed to extracurricular, Mae commits herself to rise "quickly through the PartiRank" (193). As the narrator continues, "By Thursday night, she'd gotten to 2,219, and knew she was among a group of similar strivers who were, like her, working feverishly to rise," much as religious believers rise to heaven through faith (193). Most notably, SeeChange, a revolutionary and affordable digital camera that owners can place in surreptitious locations, offers godlike omnipotence to everyday people through surveillance, as suggested by its marketing slogan: "ALL THAT HAPPENS MUST BE KNOWN" (68). As Bailey announces to a standing ovation in his Dream Friday, Ted Talk–like presentation, which unveils the new device to an audience that is enthusiastic about the prospect of being raised to the apparent height of deities, "We will become all-seeing, all knowing" (71). As he intimates by virtue of referencing Egyptian protesters who used their cell phones to broadcast the Arab Spring, humanity will see the realization of the kind of moral world that religious believers seek: a world in which perpetrators of human rights violations are exposed and punished.

This cybercapitalist fundamentalist desire to know, own, and disseminate data and to make viral the value of data and metrics creates an array of problems for Americans in the world of Eggers's novel, one of which is the spread of misleading information that masquerades as reliable. Much as the religious metaphors that fundamentalists read literally come to function as false facts that in essence negate the need for faith to exist, the data that the Circle gathers, most notably through Customer Experience, come to exist as untrustworthy and thus evocative of online information as Ferris portrays it in *To Rise Again at a Decent Hour*. In accord with her colleague Jared's training, Mae gathers reliable data that morph into unreliable information when she pushes customers with whom she communicates through her work in Customer Experience to assess and then reassess the quality of the service she provides. If the customer's score of her work returns as anything less than 100 percent—even a single point less—then Mae seeks "clarity" on the cause of the relative imperfection because, as Jared puts it, referencing the apparently collective mentality of Circle employees, "that missing point nags at us" (52). In other words, Mae resembles other Customer Experience representatives in that she nags customers. In doing so, she receives new and faulty data: perfect scores that only exist because customers feel annoyed by her follow-up questions and want them to end. She renders Customer Experience as a place with no tangible "unknowns," but the known information that Customer Experience gathers exists as a virtual reality that is disconnected from lived reality (326). Along the same lines, LuvLuv, a Circle-born dating site that houses online profiles of prospective lovers, presents faulty information about individuals who use the site. When a Circle employee named Gus unveils the new media platform in a Dream Friday presentation with the help of Francis Garaventa, a hacker-turned-Circle-employee who works to develop child safety technology for the Circle and who develops a romantic interest in Mae, Mae sees firsthand the problems with the site. As she watches Francis reveal his own affection for her publicly by way of referencing her LuvLuv profile, Mae thinks about what she sees as "some kind of mirror" that distorts her as the lovers of Kristen Roupenian's "Cat Person" distort themselves through text messages (*The Circle* 126). Mae sees the profile as a "matrix of preferences" that attempts to capture her "essence" but flattens the reality of her personhood, failing to represent her accurately (126).

A cybercapitalist fundamentalist focus on information of a digital variety likewise creates problems with conceptions of history. It renders more thorough pictures of the past that upend the social order that pre-digital-age notions of the past produce to personally devastating ends. Most notably, PastPerfect, Bailey's "passion project," which aims to "fill in the gaps in personal history" and "history generally," dismantles the sense of self that Annie has because it reveals her ancestors' transgressions and invites her to identify herself with those transgressions (410). A blueblood who knows that her ancestors came to the New World on the *Mayflower*, Annie demystifies the source of her privilege through PastPerfect, learning that it comes at the expense of social justice. Annie initially learns that her British ancestors held and sold slaves in Ireland and in the United States as well. She also learns about her own parents' indiscretions. She discovers that they had an open marriage—a revelation that alters her sense of them—and she discovers that they once failed to help a man who fell from a pier. As Annie describes the surveillance footage that PastPerfect shows her, "they just got up and left. They never called 911 or anything. There's no record of it" (443). She fails to see a distinction between her own identity and ethics and her parents' and ancestors' actions and moral compasses. To cite Mae's words, she fails to see that something that happened several years ago or "six hundred years ago has nothing to do with" her—or at least not to the extreme degree that Annie comes to imagine it does (432). As a result, by novel's end, Annie experiences a nervous breakdown, enters into a coma, and is kept alive by way of machines that are paradoxically analogous to the digital ones that brought about her emotional duress. She is kept alive as a cyborg after the trauma of her exposure to information from invasive digital media.

A desire for information of the sort that cybercapitalist fundamentalists deify, too, results in the mechanization and denaturalization of nature. Early in the novel, Eggers describes Mae's therapeutic and rejuvenating experiences of going into nature—of going off grid, so to speak, and kayaking in the San Francisco Bay with a kayak she rents from Marion Lefebvre, the owner of Maiden Voyages. The world Mae encounters on her rented kayak is textured and juxtaposes with the flat digital world. In it, Mae floats in deep waters—above layers of leopard sharks, bat rays, jellyfish, and "the occasional harbor porpoise" that swims unseen (83). She feels "calm" and "strong" in part because she rises to the challenge of exploring not a world of cybermystery as DeLillo portrays it or conspiratorial mystery as Pynchon does, but mystery as American Transcendentalist Herman Melville portrays it in *Moby-Dick* (*The Circle* 83). According to the narrator, the fact that the creatures were "hidden in the dark water, in their black parallel world" feels "strangely right" for Mae, as does "knowing they were there, but not knowing where, or really anything else" (83, 84, 83–84). However, SeeChange cameras create a metaphorical sea change in the aquatic environment. These cameras render the natural world as webbed and connected—as a demystified text, so to speak, that is akin to a sacred text as a literalist religious fundamentalist reads it. In other words, these cameras render the natural world as part and parcel of the emergent digital and globalized world that enables the Circle as a cybercapitalist fundamentalist institution to make money. Notably, when these cameras capture a digital video of Mae's act of borrowing a kayak from Maiden Voyages after hours, they create the opportunity for authorities—and for Bailey—to read the ill-advised, unwise, yet relatively innocent act as transgressive. Police arrive at the scene and treat Mae as a criminal until the sympathetic Marion ultimately opts against pressing charges against her. Mae's realization that the natural world is no longer an unwebbed enclave from the corporate, digital world dissipates her interest in kayaking altogether. She never returns to Maiden Voyages or the bay after realizing that it is, to appropriate DeLillo's words from *Underworld*, in "the grip of systems" (825).

Much as cybercapitalist fundamentalist values lead to the colonization and transformation of nature through the placement of digital devices that gather data about the natural world, they threaten the U.S. government and the ideals upon which that government is founded. Initially, and in ways evocative of the conflation between President Midwood and Eric that DeLillo portrays in *Cosmopolis*, Eggers represents American politicians as ushering in a conflation between corporations and government by wearing SeeChange cameras that allow them to *go transparent*, to reference Bailey's term for revealing one's everyday activities through wearing a SeeChange camera that morphs life and work into the stuff of reality television.

Congresswoman Olivia Santos is the first of several fictionalized politicians to agree to "show how democracy can and should be: entirely open, entirely transparent," making manifest an embodiment of "Sunshine Laws" that "give citizens access to meetings, to transcripts" and making manifest a parody of religious fundamentalist textual literalism (*The Circle* 210, 207, 207). Eventually, the Circle expands its efforts to colonize the American government by revealing a plan to utterly supplant American democracy with "*Demoxie*," "real and unfiltered—and, most crucially, complete—democracy" (400, 401). As a part of this plan, the Circle plans to register all Americans to vote and require that all citizens cast votes. It likewise plans to collect citizens' taxes through Circle accounts. As a result, Demoxie as a digital platform and idea will be able to eliminate lobbyists, polls, and Congress. As Stenton explains through a rhetorical question that suggests the monopolistic if not outright totalitarian cybercapitalist fundamentalist reality he desires, "If we can know the will of the people at any time, without filter, without misinterpretation or bastardization, wouldn't it eliminate much of Washington?" (395).

In addition to working to dismantle the American government, the Circle as a cybercapitalist fundamentalist institution works to colonize humanity with the goal of dehumanizing humans. Most notably, the Circle transforms humans into what Andy Clark calls natural-born cyborgs. It morphs humans into beings who are evocative of textual literalism—beings who, for instance, wear medical bracelets that make their bodily functions transparent to their employer and beings who broadcast the events of their everyday lives as transparent politicians do. Indeed, Bailey manipulates Mae into becoming a natural-born cyborg by going transparent. As a result of talking with Bailey, Mae believes herself to have "an awakening" to the notion that she is in dire need of salvation due to her transgression at Maiden Voyages (296). She has the "revelation" that, to quote the emergent cultic maxims she develops for an audience of Circle employees, "SECRETS ARE LIES / SHARING IS CARING / PRIVACY IS THEFT" (305). And she agrees to atone for her sin by wearing a SeeChange camera on her person at nearly all hours of the day (296). In doing so, she becomes "a window" like one that can be found in Microsoft Windows (309). And her life becomes a reality television performance, as intimated by the narrator's remark that watched individuals "perform" their "best self" (330). She becomes a walking and talking SeeChange camera advertisement and an advertisement for the Circle, touring departments and introducing products. She also becomes a fake but superficially good and constantly plugged-in, inhuman, and torn kind of human that the Circle endorses. Although she believes that the metaphorical "tear" within her is "not knowing," it more likely reflects the bifurcation of Mae's life (470). Over the course of the novel, she increasingly comes to leave behind her humanity to function as a corporate robot.

As Eggers suggests, humans who serve corporate interests and merge with digital devices or outright become equivalent to them emerge as increasingly less capable of creating meaningful as opposed to superficial, digital-age connections with one another. Most significantly, Mae loses her interest in the prospect of meaningful romantic love as she comes to function as a cog in the Circle's cybercapitalist fundamentalist system. Mae initially shows a romantic interest in a character whom Eggers positions as a human embodiment of mystery, which the Circle as a cybercapitalist fundamentalist entity seeks to demystify. She shows an interest in Kalden, who unbeknownst to her until the novel's end is secretly Ty Gospodinov, the prodigious Wise Man who develops the digital devices and media platforms that Bailey and Stenton market. In an era of social networks, Kalden, to appropriate the narrator's words, walks metaphorically "in a new direction" and "through a narrow, shadowed path, alone" (171). As a "calligraphic man," an individualist who resists conformity or superficial community in a digital age that proliferates both, he keeps details of his identity inaccessible and unwebbed, so to speak (218). Although his aura of mystery excites Mae at first, she comes to realize that her sense of wonder about Kalden is unsustainable. In the narrator's words, "Kalden was the only man for whom she's ever had real lust," but "she was finished" because she "would rather have someone lesser if that person were available, familiar, locatable" (236). When Mae fails to unearth digital information about Kalden or see him as regularly as she would like to, she instead turns her affections to Francis. Perhaps due to psychological trauma he experienced in his youth when his sisters are raped and murdered, Francis pales in comparison to Kalden as a sexual and emotional partner for Mae. Each of the sexual encounters that Mae and Francis have results in Francis's premature ejaculation, suggesting that Mae is unable to experience sexual satisfaction of the sort that Kalden can provide. Mae and Francis's sexual encounters emerge as increasingly mediated both literally and metaphorically. During their first encounter, Francis violates Mae's privacy by creating a digital recording of her without her permission. Eventually, he and Mae integrate their digital medical bracelets into their sexual activities, finding satisfaction in watching their pulse rates rise on their respective screens. Finally, Mae agrees to play out Francis's corporate, digital-age, metrics-oriented fantasy by rating his sexual performance. Even though he ejaculates prematurely, she gives him a perfect score, realizing the Circle's corporate fantasy of perfection despite the reality of imperfection.

Much as religious fundamentalism has potential to lead to violence in the form of fanatical, religiously motivated terrorism, cybercapitalist fundamentalism has potential to produce violence because it dissipates opportunities for authentic human connection. For instance, when Mae introduces SoulSearch to the masses as a technology that tracks down "fugitives worldwide" in order to "create a safer and saner world," the presentation paradoxically showcases everyday citizens as potentially insane: as militant and armed with digital devices that come to function as weapons (450). In inviting "watchers" to help her track down Fiona Highbridge, a forty-four-year-old British murder convict who escaped prison, Mae provides corporate sanction for their act of trapping Fiona with digitally recording cell phones that take on the aura of guns (454). Mae exposes the digitally connected masses' violent desire to "[l]ynch" Fiona instead of allowing her to engage in a legal, controlled, and ideally humane judiciary process (455). Hence Mae exposes that the impulse toward violence exists as ubiquitous in the digital age and in part because of digitization, which dehumanizes humanity. Along the same lines, Mae displays the violent threat that cybercapitalist fundamentalism poses through her second SoulSearch demonstration. She attempts to locate Mercer for personal purposes despite his desire to disappear into "the densest and most uninteresting forest" he can find in a northern California wilderness that he naïvely imagines as existing beyond the Circle's corporate reach (435). Despite the audience's enjoyment of the spectacle, the search for Mercer develops into what Mercer perceives as a "creepy stalking expedition" (461). Mercer attempts to escape masses of everyday people who come to function as hunters working for the Circle as well as drones that Mae releases in "a voice meant to invoke and mock some witchy villain" (462). Mae, of course, has unbeknownst to herself *become* the witchy villain that she thinks she merely mocks, as evidenced by the fact that her efforts to force Mercer to "acknowledge the incredible power" of technology lead him to a horrifying death (464). Ultimately, Mercer drives his truck into a gorge in his attempt to escape Mae, her digital drones, and the masses of digitally connected metaphorical American drones who mindlessly help Mae enact violence under the guise of a tech demonstration.

Eggers intimates that violent events such as those that manifest as a result of increasing engagement with digital technology have potential to generate a cybercapitalist fundamentalist apocalyptic reality that is similar to the kind of violent end of time that many religious fundamentalists desire. As Eggers characterizes it, however, what counts as an apocalypse in the age of globalization exists as a matter of opinion. Whereas Kalden, who reveals his true identity as Ty to Mae near novel's end, sees the completion of the Circle as a "totalitarian nightmare" and as an apocalyptic "end" in which everyone "will be tracked, cradle to grave, with no possibility of escape," Mae sees a dramatically different future as apocalyptic (486). For Mae, Kalden exists as a liar who scorned her as a lover, and in part because of her feelings of betrayal, his proposition to unweb the world and essentially abolish cybercapitalist fundamentalism horrifies her. Because she so desires to be "seen"— perhaps because of her humble roots—Ty's dream of off-grid invisibility repels her (490). His offer to "vanish" together in Asia or by sailing around the world is an apocalyptic nightmare according to the worldview that the Circle has successfully sold to Mae (491). As a result, Mae sees the reality that Ty aims to make manifest through "The Rights of Humans in a Digital Age," a statement that he writes and asks Mae to read for her digital audience, as apocalyptic (490). By advocating for human rights, Ty jeopardizes cybercapitalism. He threatens to limit the power of the Circle and keep digitization in check in ways that would render Mae less visible. Thus, she sees herself as averting an apocalypse that Ty threatens to realize by essentially choosing the Circle, a lifeless corporation, as her metaphorical lover over Ty as an actual lover. She reports Ty's plan to Bailey and Stenton, and with what Mae blindly calls "their customary compassion and vision," Bailey and Stenton realize what Eggers's authentically compassionate readers certainly view as apocalyptic for Ty (497). They allow Ty "to stay on campus, in an advisory role, with a secluded office and no specific duties" (497). They essentially imprison Ty to keep him from spreading his heretical, countercybercapitalist fundamentalist message.

Although the apocalypse as Ty and Eggers see it exists as unavoidable in the world of Eggers's novel, Eggers provides opportunities for readers to envision modes of resistance that might thwart the realization of a similar dystopian or outright apocalyptic reality in life. The primary mode of resistance that he puts forth involves putting a premium on the achievement of literacy and understanding, which DeLillo celebrates in *Cosmopolis*. Throughout the novel, Eggers invites his readers to read his text in search of acts of understanding and with an investment in attaining understanding that adds depth to superficial acts of knowing as the digital age proliferates them. Most notably, Ty's approach to understanding the workings of digital devices functions as a model to readers who seek to develop their relationships with digital devices at a technologically saturated moment in history. Despite his role as a founder of the Circle, Ty resists the capitalist system that flattens digital devices into corporate tools because he sees digital devices and media as more than mere products. He sees depth in the demystified, cybercapitalist fundamentalist flat world because he has the intellectual capacity to see beyond the superficial surface of a digital device. In other words, he understands the science of how seemingly mysterious digital devices work, as evidenced by the narrator's description of him in an encounter with Mae at her screen-ridden workstation early in the novel. As the narrator explains, he looks at one of her flat screens, "but his eyes were seeing something deep within" (94). Similarly, Ty sees literal depth in the Circle as a corporation because digital technology that is free from the snares of capitalism is what most interests him. In what Eggers portrays as a mystery-rich, deeply romantic, and emotionally meaningful experience, Ty guides Mae into the depths of the Circle, where massive computers give physical texture to the apparently flat world of digital data. Ty even shows her and Eggers's readers that humans can humanize machines instead of being mechanized by them by introducing Mae to a device that he has humanized. He introduces her to Stewart, a data storage unit, and to other similar units that "hum" with life (222). They hum in a way that runs counter to the deadening hum of digitization that Egan critiques in *A Visit from the Goon Squad.*

Eggers, too, suggests that reading for allegories and metaphors equips contemporary citizens to counter cybercapitalist fundamentalist efforts to demystify and commodify the increasingly digital world. Eggers aims to help equip his readers to engage in resistance of cybercapitalist fundamentalism by employing literary devices that come into conversation with the overarching metaphor he employs in his novel: that of multinational corporations existing as religious institutions in the ever-globalizing world. For instance, as Margaret-Anne Hutton points out, Eggers alludes, in his novel, to the allegory of the cave from book 7 of Plato's *Republic* when he portrays Ty as guiding Mae—and Eggers's readers—from the data storage room to "a great cave, thirty feet high, with a barrel-vaulted ceiling," a cave where Ty lives that "was supposed to be part of the subway" but now sits "empty, a strange combination of manmade tunnel and actual cave" (*The Circle* 223). As Hutton sees it, Eggers's allusion aims to prompt readers to reconsider the concept of reality in an era in which reality "is saturated with screens" (Hutton 193). It strives to invite them to see a metaphorical connection between Plato's allegory as a meditation on the relationship between reality and illusion and Eggers's meditation on the same subject, albeit in a twenty-first-century context. Plato's text, which presents a dialogue between Glaucon, Plato's brother, and Socrates, Glaucon's mentor, endeavors to reveal the way in which individuals chained to a wall in a cave falsely perceive shadows reflected on that wall as reality. In his narrative, Plato suggests that the enslaved individuals he describes can, by breaking free and seeing the world beyond the cave, equip themselves with the knowledge they need to understand the false reality that the shadows in the cave present. As a result, they can serve as effective philosophical rulers of the ideal political community that *The Republic* on the whole describes. By contrast, in Eggers's text, Mae and Eggers's readers alike travel away from a world of illusion at the surface of the Circle's campus and from a real twenty-first-century world that is saturated by digital media and virtual reality. They travel into Ty's hybrid cave, which demystifies mysteries of human connection for them and which thereby functions as an inverse of Plato's cave. Although neither Mae nor Eggers's readers necessarily emerge as philosophers of the kind that Plato describes, Eggers's readers ideally emerge, from the experience of reading, as better equipped to understand the difference between superficial digital and authentic human connections, the latter of which Mae and Ty develop.

They emerge with a foundational piece of a philosophy of technology that Eggers appears to believe all twenty-first-century Americans need.

Likewise, Eggers includes allegories and metaphors of his own creation with the goal of having readers engage with counterliteralist and thereby countercybercapitalist fundamentalist ways of thinking. Most notably, he portrays an aquarium that takes on metaphorical and allegorical significance as it leads readers to better understand the workings of multinational corporations in the globalized world. As Eggers describes it, the vast aquarium contains numerous sea creatures that Stenton brings back to the Circle with him from an expedition to the Marianas Trench, which is located in the western Pacific and which contains the world's deepest natural trench, the opposite of the flat and superficial world that the Circle produces. The star creatures on exhibit in the aquarium include Stenton's prized catch: a shark that the narrator describes as "ghostlike, vaguely menacing and never still" and that has "translucent skin, which allowed an unfettered view into its digestive process" (*The Circle* 309, 310). On exhibit, too, is a "pale spineless" octopus that behaves "like a near-blind man fumbling for his glasses"; that seems to "want to know everything" such as "the shape of the glass, the topography of the coral below, the feel of the water all around"; and that alternates between apparent confidence and utter weakness, as demonstrated by the fact that its "shape seemed to change continuously" (311). Finally, the aquarium showcases a "fragile," "exhausted," and "shy" seahorse with "intelligent eyes" and no defenses—a beautiful creature that hides in the aquarium and fails to realize that Circle devotees watch him vigilantly "on the other side of the glass" and can "see everything" (315). Metaphorically, as the narrator's description of the seahorse's circumstances in particular suggests, the aquarium on the whole functions as the Circle's digitally connected world does: as connected for the purpose of surveillance. Just as Mae and the watchers who tune in to her broadcast peer into the aquarium to assess and further denaturalize elements of the formerly natural world, Stenton and Bailey create a world that they can watch and subsequently manipulate because of information they acquire. They create a world that resembles the aquarium: a world in which no one is free despite the enduring and ever-American illusion of freedom.

Yet Eggers layers allegories and metaphors in his novel in order to exercise the critical imaginations of readers who benefit from engagement with and the experience of understanding literature in the face of cybercapitalist fundamentalism. Indeed, readers are invited to interpret the three creatures that Eggers spotlights as allegorical representations of the Circle's three Wise Men. They might see the vicious shark as an allegorical representation of Stenton, the most profiteering and colonial of the Wise Men. They could interpret the spineless octopus as the equivalent of Bailey, who does Stenton's bidding by serving as the Circle's appealing face and rhetorically savvy mouthpiece—a rhetor who consistently and falsely draws connections between the Circle's work and social justice work in the world. And they might see the intelligent yet fearful seahorse as resembling Ty, the metaphorical brains behind all the Circle's technology who has a "nose like some small sea creature's delicate snout" and who, despite his central role in the Circle, essentially lives life while hiding in subterranean regions of the campus (167). In turn, they might see allegorical significance in the octopus's and seahorse's inevitably violent fates beyond the allegorical significance toward which Ty gestures when he observes that the Circle as a monopolistic corporation is the "fucking shark that eats the world" (484). When a Circle employee named Victor disrupts the manmade ecosystem that the aquarium constitutes by inviting the shark to join the other creatures for the purpose of creating an authorized spectacle for the masses, predictable and noteworthy violence ensues. To a look of "fascination and pride" by Stenton, who corresponds allegorically to the shark, the shark eats both the octopus and the seahorse (481). Hence the scene suggests that Stenton eats Bailey and Ty as cofounders—or, rather, that his corporate values overpower their values entirely. In addition, the scene showcases through metaphor the problem with what Mae characterizes as "the natural thing to happen" (479). Because the natural thing is devastating, the scene illustrates the need for manmade protections that safeguard elements of nature that are always already ecologically entwined with the manmade, technological world. In other words, the scene suggests that nature and the natural order of things are vulnerable and that the American government can and should function to keep corporations such as the Circle in check, with the goal of protecting elements of nature that Americans value.

Moreover, Eggers complements metaphor and allegory with references to real and fictionalized works of art and artists that resemble those that DeLillo makes to Rothko and Brutha Fez in *Cosmopolis*. These references invite readers to contemplate and better understand the function of art in the face of cybercapitalist fundamentalist, data-driven literalism. Real works of art that Eggers mentions in his novel include, for instance, the Romanian Constantin Brâncuși's apparently abstract sculptures, which Brâncuși saw as realistic for the ways in which they captured the true and inevitably metaphorical essence of things. They likewise include the American Donald Judd's minimalist sculptures, which were often made of industrial materials that were evocative of human technological progress and that Judd saw as literalist because they referenced nothing in the world beyond themselves. As Roberta Smith puts it in a 1995 *New York Times* article, Judd seemed "determined to prove" that art "wasn't only esthetic" but "that there were broader, more practical applications for its underlying principles." The narrator references

Brâncuși's work when Mae first sees Ty's real-life face, which resembles a Brâncuși sculpture because it is "smooth, perfectly oval" (Eggers, *The Circle* 167). And Mae references Judd's work. Armed with some knowledge of art because she considered completing an art history major in college, she mentions Judd in a conversation she has with Ty, who saw Judd as "a big inspiration" to him in his creation of Stewart as a data storage unit and inevitably an aesthetic object (220). As Ty explains to Mae, whose knowledge of Judd is limited to what little she remembers from one college course, he in particular appreciates Judd's perspective on reality: the notion that, according to Ty's quotation of Judd's oft-quoted words, "[t]hings that exist exist, and everything is on their side" (220). As Eggers intimates by making reference to both Brâncuși and Judd and to their divergent philosophies, the relationship between art, reality, and social purpose is complex. Art exists as part of reality while expanding human notions of what counts as real. It, too, exists as capable of rendering truth beyond the relatively limited scope of reality in a conventional sense. And it exists as an effective means by which to comment on social issues and influence the real world regardless of artistic intentions that artists articulate.

Along the same lines, Eggers fictionalizes works of art within his novel as a work of art with the goal of emphasizing the potential social power of art as a metaphorical mode of expression. For instance, Eggers fictionalizes a poorly designed portrait of the Wise Men that hangs in the Circle's headquarters. As the narrator explains, it looks like "the kind of thing a high school artist might produce" because it represents Ty's, Bailey's, and Stenton's personalities "cartoonishly" (19). The narrator leaves readers to contemplate whether the painting functions as a purposeful criticism or outright mockery of the Wise Men or whether, as Annie suggests, it merely constitutes "bad art" that appeals primarily because it is "hilarious" (26). More significantly, Eggers invites readers to imagine, assess existing interpretations of, understand, and interpret for themselves a fictionalized work by a Chinese artist titled *Reaching through for the Good of Humankind*, a fourteen-foot-tall sculpture commissioned by Bailey that consists of a hand reaching through a screen and that is "made of a thin and perfectly translucent form of plexiglass" (348). Mae shows the piece, which is on exhibit in the western part of the Circle's campus, to an audience of watchers who track her movements around the campus through the SeeChange camera she wears. And in describing the piece to those watchers, she observes that the piece is "representational" even though "most of the artist's previous work had been conceptual" (348). In other words, she suggests that an audience can clearly interpret the intention of the work of art because it clearly represents what it claims to represent—even though the artist refuses to comment on the piece's meaning, preferring to allow the sculpture to "speak for itself" (349). In accord with Mae's view, Circlers whom Mae interviews understand the sculpture as representational. For instance, one named Gino suggests that the artist is "trying to say that we need more ways to reach through the screen" (348). Another Circler named Rinku suggests that she sees the screen as a "barrier" that "the hand is transcending" with the goal of making the "connection" between the Circle and its "users stronger" (349). Eggers, however, hopes that his readers come to realize that the sculpture functions as conceptual art because the idea it represents is its most important element. He hopes his readers come to see the work as critiquing the Circle's invasive and threatening nature. Hence he hopes his readers understand the work as sustaining a counterliteralist, countercybercapitalist fundamentalist purpose.

Ultimately, Eggers's *The Circle* functions much as *Reaching through for the Good of Humankind* does in that it exists as conceptual art that invites readers to develop their own identities. To quote Ty, Mae is "in a unique position to influence very crucial historic events" at a "moment where history pivots," and, similarly, Eggers sees his readers as uniquely positioned to influence the future that manifests in the real world, which gives context to the literary digital dystopian novel that *The Circle* constitutes (405). Much as DeLillo sees readers of literary fiction as critical thinkers who can shape the future, Eggers sees his readers as positioned, at an important historical moment, to better parse fact and fiction through the act of reading about truth and lies in a humanitarian work of fiction. In other words, Eggers sees his readers as, through the act of reading fiction, positioning themselves to parse rhetorical claims of the sort that Bailey as the Circle's ambassador makes about the Circle's promulgation of human rights abuses that masquerade as social justice initiatives. As Eggers suggests, readers can interrogate and analyze analogous rhetoric that they encounter in the real-life globalized world in which they live—a world that teeters on the brink of dystopia. They can attain deep understanding that runs counter to the metaphorically flat knowledge that ubiquitous digital data proliferate in the so-called flat world. Based on the deep senses of understanding of the world and social justice issues that readers attain, readers can make meaningful connections with one another that enable them to act responsibly as activists and advocates for change who keep multinational corporations such as the Circle in check. They, too, can limit use of digital devices that threaten to disconnect individuals from one another and outright dehumanize humanity.

In addition to inviting readers to develop their identities as socially responsible citizens through acts of reading, Eggers invites readers to see *The Circle* as an example of how twenty-first-century authors can revitalize literature and thus develop literature's identity in the digital age. Much as information may never "prevent abuses of power" as Bailey dreams it will, literature will never prevent abuses or even atrocities (66). Yet literature manages to avoid irrelevance and stagnation despite widespread claims such as those that Sven Birkerts makes in *The Gutenberg Elegies*. It manages to remain relevant in the face of digital media that consistently threaten but fail to render it as less consequential. Indeed, literature continues to help humanity picture human rights abuses and possibilities for social justice in the globalized world. It continues to help humanity play out scenarios to their logical ends and also to imagine the future in both utopian and dystopian senses. In turn, it helps humanity imagine a middle ground between the extremes that utopian and dystopian works represent. It helps twenty-first-century citizens picture some realistic and realizable reality that eludes the corrupt world of Eggers's novel and the problematic world that Eric helps create in DeLillo's *Cosmopolis*: a future in which humanity relies on digital devices and philosophies of technology in equal measure. Hence socially concerned literature such as DeLillo's and Eggers's disseminates a noteworthy desire that runs counter to Mae's in the closing pages of *The Circle*. Whereas Mae sees the Circle's metaphorical final frontier as mind reading, as evidenced by the way she ponders over what might be "going on in" the comatose Annie's head, authors of socially conscious literature see possibilities for an alternate manifest destiny, so to speak (497). They seek to expand human minds instead of reading them. They see that literature remains one of humanity's best tools for producing an open, educated, and understanding civilization—one that offers riches far more valuable than those that multinational corporations seek to accrue through marketing devices that rob humans of their rights and threaten their ability to think for themselves about when, why, and how to integrate technology into human experience.

CHAPTER 5

National Divides and Digitization in Zadie Smith's "Meet the President!" and Mohsin Hamid's *Exit West*

Cybercapitalism as both Don DeLillo and Dave Eggers describe it in their respective novels contributes to the perpetuation of globalization, which Manfred B. Steger defines as the "intensification and stretching of economic connections across the globe" and thus the stretching of these connections across national divides that set pre-digital-age terms for identity (37). Globalization influences political, cultural, and social features of everyday life, as evidenced by Thomas L. Friedman's utopian conception of it in *The World Is Flat* and by Ian Bremmer's dystopian one in *Us vs. Them*. Unlike the idealistic Friedman, Bremmer sees globalism, the "belief that the interdependence that created globalization is a good thing," as producing "ongoing political, economic, and technological changes around the world"—changes that are akin to those that Friedman outlines in his book (8–9, 6). Yet Bremmer argues that globalism and globalization widen divisions among people instead of leveling the playing field and making the world a smaller place. He argues that they invigorate populists such as American president Donald Trump, who relies on a quintessentially divisive *us vs. them* rhetoric. Hence for Bremmer, globalism and globalization produce "waves of winners and losers": socially more secure, xenophobic, and increasingly nationalistic insiders to states and less secure outsiders who exist as such because of their "racial, ethnic, linguistic, and religious" identities (6, 20).

The effects of globalization, too, influence the study of the United States and American literature, a notion that Brian T. Edwards and Dilip Parameshwar Gaonkar gesture toward in their introduction to *Globalizing American Studies*, a collection of essays that sets out to study America "as a new phenomenon" after the end of the American Century and the consequential end to notions of American exceptionalism (6). Based on what Edwards and Gaonkar argue, what counts as American literature inevitably changes as political notions of the United States change. As Wai Chee Dimock writes in her introduction to *Shades of the Planet: American Literature as World Literature*, "What exactly is 'American literature'? Is it a sovereign domain, self-sustained and self-governing, integral as a body of evidence? Or is it less autonomous than that, not altogether freestanding, but more like a municipality: a second-tier phenomenon resting on a platform preceding it and encompassing it, and dependent on the latter for its infrastructure, its support network, its very existence as a subsidiary unit?" (1). Stacey Olster makes a similar argument to Dimock in *The Cambridge Introduction to Contemporary American Fiction*, observing that American literature "resists categorization by way of authorial birthplace or citizenship or residence" because "too many writers have been born in one place and raised in or emigrated to another" (3). As a result, American literature emerges as a de facto diverse literature and a literature of diversity, a collection of texts that speaks to what it means to have been, be, become, or desire to become an American in contemporary, digitally interconnected, and globalized times. It emerges as literature about the tensions between digital borderlessness and xenophobic political efforts to establish and reinforce borders that attempt to oppress and keep out Others who are often hybrid in accord with Donna Haraway's or Homi K. Bhabha's respective theorizations of the term.

This chapter considers Zadie Smith and Mohsin Hamid as international authors of American literature of an unconventional variety. I read them as examining the United States as interconnected with other nations through globalization in their fictional works. I also read them as addressing the tensions between national and international haves and have-nots; global efforts to digitize and connect apparently disparate individuals; and efforts to disenfranchise racial, ethnic, religious, and underprivileged Others in order to reserve connected states of legitimacy for the privileged few. In the first part of this chapter, I focus on Smith, who divides her time between London and New York City (where she teaches creative writing at New York University and where she previously taught at Columbia University) and who has developed a noteworthy interest in the subject of globalization because it produces hybridity. In *White Teeth*, Smith primarily considers hybridity in the way that Bhabha theorizes it in *The Location of Culture*: as a postcolonial phenomenon that pertains to individual and national identity. She portrays characters such as Clara Bowden, the child of a Jehovah's Witness who hails "from Lambeth (via Jamaica)," and Samad Iqbal, a Bengali Muslim immigrant to England, both of whom must navigate intersections among their racial, ethnic, religious, and national identities (*White Teeth* 23). By contrast, in more recent works of fiction, among them *On Beauty* and *NW*, Smith expands her consideration of hybridity to include the peculiar effects of digital technology that allows users to have hybrid life experiences which move between virtual and material reality.[1] In writing *On Beauty* as a parody of E. M. Forster's *Howards End* that takes place in the Boston area, Smith imagines the twenty-first-century implications of "Only connect!", the epigraph to Forster's 1910 novel. And the novel allows her to reflect on her time as a Radcliffe Institute Fellow at Cambridge from 2002 to 2003,[2] the noteworthy historical moment when Mark Zuckerberg, then a sophomore at Harvard University, invented Facebook (then FaceMash), which Smith critiques in "Generation Why?", a review of David Fincher's *The Social Network*. For Smith, Facebook and other media promise social progress and transcendence but deliver only a substandard human experience. As Smith articulates it from what she calls her "Person 1.0" perspective, "When a human being becomes a set of data on a website like Facebook, he or she is reduced. Everything shrinks. Individual character. Friendships. Language. Sensibility. In a way it's a transcendent experience: we lose our bodies, our messy feelings, our desires, our fears" ("Generation Why?"). But, she implies, no actual or remotely meaningful transcendence occurs.

This analysis considers the interface among virtual, physical, and spiritual realities in the globalized, digital age as it manifests in Smith's "Meet the President!", a short story published in the *New Yorker* in 2013. Building on Anique Kruger's argument that Smith experiments "with the possibilities that arise when fiction is used to image networks, connections, and communities in a globalised and multicultural world" and also building on Benjamin Bergholtz's consideration of Smith's treatment of globalization in her fiction,[3] I suggest that "Meet the President!" provides a commentary on how the United States helped produce the present digital and globalized age, functioning as a fictional counterpart to "Generation Why?" (Kruger 69). In the story, Smith portrays fourteen-year-old protagonist Bill Peek's interactions with Melinda Durham and the eight-year-old orphan Agatha "Aggie" Hanwell, fictionalized lower-class locals who live in 2053 in Felixstowe, England, a coastal town that Peek visits with his father and the former real-life home of Smith's father Harvey. As an Incipio Security Group executive, Bill's father gives his son access to an AG 12, a digital device that Bill wears to play Blood Head 4, a violent game that alters his sense of reality by inviting him to traverse a virtual post-apocalyptic Washington, DC, landscape with the goal of saving and meeting the American president. I argue that in telling Bill's story of moving between virtual and real life and between England and America, a nation of apparently increasing interest to Smith,[4] Smith provides an international perspective on digitization. She creates a literary philosophy of technology that draws attention to ways in which digital technology redefines conceptions of local, national, and global space and, in turn, recasts notions of humanity, human interconnection, and the natural world as they exist in that space. In other words, Smith explores the political incarnations and implications of what Sherry Turkle refers to as the "technologically enmeshed relationships" of our contemporary moment—relationships that Turkle says "oblige us to ask to what extent we ourselves have become cyborgs, transgressive mixtures of biology, technology, and code" who resemble the natural-born cyborgs that Andy Clark theorizes (*Life on the Screen* 21). Ultimately, the disparities that define Bill's virtual and real-life experiences provide a metaphorical peek into the very real problems of technology that Smith believes digitally connected twenty-first-century citizens face. As Smith complements the virtual apocalyptic and postapocalyptic conditions of Bill's virtual reality with apocalypse as revelation, Bill has the opportunity to understand his own physical nature, mortality, and social circumstances. In turn, Smith's readers have an opportunity to see the spiritual detriments of human disconnection in the seemingly borderless digital age. Through engagement with literature, they might begin to imagine alternatives to spiritual depravity and realize a more socially just future than the one that humanity is on course to encounter.

In the second part of this chapter, I focus on Hamid, a self-proclaimed modern-day nomad who has called Pakistan, England, and the United States home and who writes from what critics have characterized as a liminal or deterritorialized position.[5] Like Smith, Hamid has witnessed the world transform from a state of relative disconnection to globalized connection. In describing the experience of moving from California to Lahore at the age of nine, Hamid observes that in "1980 there were no email accounts or social media or text messages," so moving made him feel as though he had "left one world and entered another" ("Mohsin Hamid on the Dangers"). He had no good way to stay in touch with friends or to screen American culture since Lahore at the time had "only one television channel, broadcast for only part of the day, with only one or two shows a week that [he] felt any desire to watch." By contrast, in the globalized and networked twenty-first-century present moment during which Hamid writes, he can board "daily flights from Lahore to Rio de Janeiro, to Sydney" and he can also readily "travel the world" by "phone and computer." Moreover, he can travel through time, using social media to "sift endlessly through these archives of past moments," and he can "commingle them with present choices and likes and filters" to "craft new past-present hybrids." He can live a hybrid life, to stretch Bhabha's notion of hybridity, traversing time and geographical space with the help of digital screens despite the boundaries and borders—screens of a different kind that divide people from one another—that nationalism, xenophobia, and state- and non-state-sanctioned terrorism work together to create.

Emerging out of Zygmunt Bauman's exploration of mass migration and the ways in which media represent and perpetuate problems involving it in *Strangers at Our Door*,[6] my analysis of Hamid considers the relationship between national borders that function as divisive screens and screen-based digital technology that speaks to the problems and possibilities of globalization in *Exit West*. Published at a noteworthy historical moment shaped by events such as Brexit and U.S. president Donald Trump's promises of deportation and a border wall, Hamid's novel, like Smith's story, offers an international perspective on the politics of digitization. It attends to changing realities of the digital divide and addresses human interconnection in both content and form.[7] It stretches the concept of connectivity beyond digital contexts, telling a range of stories about human connections and disconnections through vignettes and through the main narrative of Saeed and Nadia, tech-savvy lovers from an unnamed, war-torn city in the East who exit west out of it into impoverished refugee status through magical doors that lead to Mykonos, Greece; to London, England; and, eventually, to the San Francisco Bay Area, the home of Silicon Valley and digital technology. I argue that as refugees in digital times, Saeed and Nadia continually find themselves "present without presence" or *alone together*, to reference the third work of Sherry Turkle's trilogy on computers and people (Hamid, *Exit West* 40). In other words, the presence of enchanting digital technology that connects users while also dividing their attention, the presence of alienating state-sanctioned and terrorist violence, and the absence of state-sanctioned presence for undocumented immigrants in xenophobic nations that seek to screen them out present them with a paradox of existence. They live as simultaneously connected to and yet disconnected from each other, their homes, and the nations to which they migrate. And they live divided among past, present, and future while tethered to physical screens that Hamid aestheticizes because he sees digital art and art about digitization as poised to reshape future conceptions and uses of screens, be they borders or digital objects. Ultimately, Hamid suggests that works such as Thierry Cohen's digital photography or his own novel afford perspectives that mass-consumed and arguably toxic digital media do not. And through perspectives that they attain via art, Hamid's readers can contemplate a future of fruitful interconnection through digital and nondigital means that displaces the disconnection which borders perpetuate, not people. They might come to see the possibilities that interconnection and hybridity afford and come together in a common hope for a more socially just future, even if citizens and nations such as the United States are not yet ready to realize that future.

Digital Dystopia and Possibilities for Human Redemption in Zadie Smith's "Meet the President!"

In "Meet the President!" (2013), Smith draws attention to the power digital technology has to pervade human or posthuman consciousness and to redefine conceptions of global reality.[8] She portrays Bangkok-born Bill Peek as a symbolic product of the globalized times, observing that he sees himself as "simply global" as he jaunts between locations such as Japan, Norway, Mexico, and England, "accompanying his father on his inspections" of digital, military products developed by his multinational corporate employer, Incipio Security Group ("Meet the President!"). As a global citizen, Bill is a cyborg who is not so much a mix of man and machine but a representation of the interplay between humanity and cybercapitalist ways of thinking and being that authors such as DeLillo and Eggers explore in their respective novels about digitization. Evidence of Bill's existence as a global citizen emerges, for instance, in the way in which he sees landscapes from across the globe and from across time: as readily accessible to him. As he steps onto the town's beach at low tide, he thinks not about the limits that geographical features such as seas or other borders create. Instead, he imagines possibilities for transcending borders because globalization purportedly allows for such transcendence in other contexts. According to the narrator, it seems to Bill that he can "walk to Holland." Moreover, because of the digital technology to which Bill has access through his father's company, he perceives the world as globalized, and hence he sees sites from around the world as ever accessible. While walking across Smith's futuristic representation of Felixstowe over the course of the story, Bill is able to experience a virtual futuristic fiction within Smith's literary futuristic fiction. The unregistered AG 12 device that Bill's father's company produces for military purposes and that his father gives him "as a bribe and a sop" to leave Tokyo and come to England to be with him allows Bill to obtain information about his real-life surroundings and play Blood Head 4, a violent digital game that lets its players traverse the globe in accord with globalization's ideals by selecting the nation in which they want to play.

According to Smith, digital devices have a religious aura in the globalized world as they do in DeLillo's *Cosmopolis* and Eggers's *The Circle* because humanity so relies on them for connectivity. And these devices also speak to local and national histories—both religious and secular—that have shaped the globalized present. When Smith's narrator first describes Bill using his AG 12, which functions as his *A*rtificial *G*od (to present a possible explication of Smith's acronym), she observes that a light evocative of a halo encircles his head. This image suggests that Bill exists as a divine being while he uses the device or at least that he sees himself as divine because of the power with which digital technology imbues him. Bill's metaphorical sanctification or deification points to what Frank J. Lechner and John Boli see as a connection between globalization and religion: the notion that religious believers attempting to convert others to their faiths functioned as the world's first globalizers (Lechner and Boli 387). Bill's sanctification or deification in this image likewise speaks to the religious rhetoric that Smith employs in her description of Bill when he begins playing Blood Head 4 in his selected city of Washington, DC. Notably, the narrator evokes John Winthrop's "A Model of Christian Charity" and also Jesus's words from the Sermon on the Mount in Matthew 5:14 in observing that "[a]nother world began to construct itself around Bill Peek, a shining city on a hill"—or at least a postapocalyptic parody of the American national and religious metaphor ("Meet the President!").⁹ In Blood Head 4, the United States has fallen from its position of prominence, perhaps because of imperialist ambitions as they manifest through globalization. It no longer functions as the shining example to the world that Winthrop imagined it would be and that it perhaps in ways became during what Henry Luce deemed the American Century. Instead, in this quintessentially fallen city upon a hill, a minotaur sits "in the lap of stony Abe Lincoln," the Lincoln Memorial; "a dozen carefully planted I.E.D.s" cover the monument; and the Washington Monument is a pile of rubble because it has been "pounded by enemy aircraft."

Smith connects this virtual, fictional, futuristic world with England's national and local history as it precedes and develops alongside and in relation to nineteenth- and twentieth-century American history, which culminates in the rise of global capitalism and in an everyday reliance on digital technology around the globe. In particular, Smith draws attention to ways in which the now-depressed and metaphorically fallen city of Felixstowe resembles Washington, DC, as the fallen city upon a hill of Bill's game. In following Bill through the futuristic Felixstowe's city streets, Smith's readers see a Martello tower, a relic of pre-rifle-artillery warfare of the nineteenth century, and they also see the ruined Felixstowe Pier, which was partially destroyed in World War II to prevent invasion and never rebuilt. They even see evidence of local history that reflects global changes which transpired because of the rise of global capitalism. They see ways in which nations with neoliberal values devastate local aspects that they contain. Specifically, Smith alludes to the North Sea Flood of 1953, which precedes other fictionalized floods that were caused by capitalist impulses that fuel global warming, a subject Smith addresses in "Elegy for a Country's Seasons." These floods, in Smith's fiction although not in real history, cause the town's residents to retreat "three miles inland and up a hill," and as a result, the town becomes yet another parody of the United States in general or Washington, DC, in particular as a city upon a hill ("Meet the President!").

Smith posits that digital devices such as the one she represents in her story foster a hybrid reality for global individuals who are perhaps drawn to them because of their familiarity with disconnection from places in real life, but not necessarily to positive ends. From the story's start, Bill experiences the attraction of his digital device and the fictional and American narrative that it imposes, and this pull results in a hybrid life experience. In meeting Aggie and Melinda, an old woman who is escorting Aggie to Aggie's twelve-year-old sister Maud's "laying out" at St. Jude's, Bill meets quintessential locals from impoverished backgrounds who stand in stark contrast to him as a privileged individual who is armed with digital technology. The encounter showcases Smith's interest in addressing distinctions between disenfranchised everyday humans who lack access to technology and privileged cyborglike characters such as Bill. Instead of showing interest in the local culture that Melinda and Aggie represent, Bill feels interrupted by them, and he initially tries to ignore Melinda as he stands "at the end of a ruined pier, believing himself quite alone" after a week of "hoping for a clear day to try out the new technology—not new to the world, but new to the boy." He moves in and out of conversation with the two throughout their interaction to realize a hybrid reality that blends his virtual reality with the real world around him. For instance, when Aggie tugs on Bill's "actual leg," she forces him out of his virtual reality and into physical reality as he presses "mute for a moment" to listen to her question. But Bill eventually reenters virtual reality, though only to find "his interest fading once again"—after he searches with his digital device for information about the sandworms he sees. He is often "impatient to return" from real lived experience to virtual experience, and finally, he "split[s] the visuals" to encompass the virtual and the real in his line of vision, coming to terms with the fact that he prefers a hybrid reality to a real or whole one. He prefers to live in the real world without paying attention to it. And he would perhaps prefer to abandon material reality altogether for virtual reality if given the option.

Smith suggests that opting for hybrid realities as global individuals opt for them involves living life not as connected, to employ a term that Smith sees as shaping digital-age thinking,[10] but as paradoxically disconnected. In one sense, evidence of disconnection manifests in relation to Bill's body, for instance when the narrator observes that Bill feels "the shocking touch of a hand on his own flesh" or when she observes that Bill was "unused to proximity" to others ("Meet the President!"). However, evidence of disconnection also manifests between people and the nations and histories that have helped shape human identity—histories for which characters in Thomas Pynchon's *Bleeding Edge* and Jennifer Egan's *A Visit from the Goon Squad* feel nostalgic. As the narrator of the story indicates, Bill's "hair and eyes and skin and name" suggest that his national heritage is English, and Melinda echoes the narrator's sense of the Peeks' history in a remark she makes ("Meet the President!"). As she explains after inquiring about her new acquaintance's personal details, there have been Peeks in Anglia for "a long time." Yet the narrator indicates that the quintessentially global father and son care little about their British roots. In the narrator's words, this topic was not "likely to engage [Bill's] father, and the boy himself had never felt any need or desire to pursue it" either. Likewise, when Bill uses his AG 12 device to pull up a history of Felixstowe and its decline over the years due to being "serially flooded, mostly abandoned," the narrator notes that he "did not care much for history" in general or, in turn, for the ways in which history shapes nations or the present moment. Bill resembles the multinational corporation for which his father works—a corporation that invents cutting-edge, future-oriented digital technology and aims to break down national divides to attain its capitalist goals. He focuses on the future. And like this multinational corporation, which is from everywhere and thus paradoxically nationless and from nowhere, Bill is from "nowhere," to cite Aggie's word. He even operates according to the logic of a multinational corporation. Since the age of six months, he has attended the Pathways Global Institute, a school located in Paris, New York, Shanghai, Nairobi, Jerusalem, and Tokyo, among other cities. And he adheres to the school's notoriously corporate and (in spirit) American motto, believing, as the school does, that above all else, above preserving or connecting with histories or nations, "[c]apital must flow."

Smith juxtaposes local culture with globalization as nations and corporations alike help to propagate it, global-minded individuals, and the digital technology that helps them thrive. She characterizes the global and the digital as a threat to humanity and humane impulses, which, ideally, national governments should exhibit. In large part, this threat manifests in Smith's text as ill will or outright hostility toward local people and things. In representing Bill's interaction with Melinda and Aggie, Smith underscores the degree to which Bill goes against his dead mother's apparently egalitarian grain by dehumanizing the two for being local as opposed to global. Although Bill's mother, who was "famed for her patience with locals," would have "tried looking the females directly in their dull brown eyes," Bill has no motivation to behave with his mother's empathy because his "mother was long dead" and "he had never known her." Instead, he sees these locals as "typically stunted, dim." He even seems to conceive of the word *local* as synonymous with ignorant, as evidenced by the narrator's remark that he sees Melinda and Aggie as "too local even to understand the implied threat" of drones that fly overhead and threaten violence over the Felixstowe beach—drones that his father's company makes and that his father is in England to inspect. Moreover, he echoes the language with which Aggie describes him but employs a more derogatory tone, observing that he believes Felixstowe "is nowhere" and stating that if "you can't move, you're no one from nowhere." He sees England as "a sodden dump" and believes that the "only people left in England were the ones who couldn't leave." Eventually, in using his AG 12 device, which Melinda sees as evidence of the fact that he is "somebody" because "they don't give 'em to nobody," Bill dehumanizes locals in a literal way through the virtual reality he projects.

Interactions between local and global characters in Smith's story spotlight the way in which global capitalism takes advantage of local landscapes and people in the digital age. Most notably, Bill as a global citizen takes advantage of Aggie as a local by virtue of using his exchange with her for personal gain. After Melinda abandons Aggie with Bill en route to Maud's wake, Bill reluctantly agrees to walk her to it, but mainly because he can project her into the landscape of his virtual reality in ways that bolster his ego as he pursues his goal of meeting the American president. He initially sees Aggie's body in virtual reality as "sometimes a dog, sometimes a droid, sometimes a huddle of rats." And he eventually transforms Aggie into "a sleek reddish fox" that he names Mystus and that he sees as functioning as a "sidekick" who "mutely admire[s]" him. He pays little attention to his sidekick and for the most part fails to empathize with her tragic situation, save in a moment when she fails to question a lie he tells: when she indicates that she believes Bill has the highest level of security clearance at his father's company. Notably, Bill capitalizes on this moment, dehumanizing Aggie by rendering her as merely a means by which to achieve his goal of completing the Global Pathways Institute's "Module 19," which emphasizes "empathy for the dispossessed." As he paradoxically touches Aggie's face with the appearance of care but with only selfish benefits in mind, his AG 12 records the incident as a commodity—as video evidence of Bill's capacity for empathy for administrators at his school. As the narrator notes in language that is evocative of religious rhetoric in Eggers's *The Circle*, Bill sees himself as "the first prophet of some monotheistic religion, bestowing his blessing on a recent convert" in this moment. By contrast, readers likely see him as subhuman in the interaction for posing as empathetic and exploiting Aggie's situation.

Moreover, Smith demonstrates the ways in which globalization and digital technology have the capacity to promote violence and terrorism of the sort that have shaped the twenty-first century, for instance because of events such as the 9/11 terrorist attacks, which ushered in the third millennium and which Pynchon and Egan in particular address in their novels about digitization. The potential for violence looms large in Smith's story as numerous drones "directed by unseen hands" fly along the coast. The drones blur the boundary between man and machine as they track, target, and kill individuals. And they speak to other ambiguities in Smith's text as well. Certainly, their relationship to AG 12 devices such as Bill's remains unclear. Readers never wholly understand whether an AG 12 has the capacity to control a drone. These drones exist as controversial, as demonstrated by the narrator's remark that Bill has learned to "despair of the type of people who spread misinformation about the Program" and as demonstrated by the fact that Bill comes to realize "those with bad intent on occasion happen to stand beside the good, the innocent, or the underaged." According to Smith, an Incipio drone tracks Aggie after she leaves the pier with Bill, and another one appears to have been responsible for Maud's death according to a local in attendance at Maud's wake. In the words of the unnamed local, "[t]hey took her from the sky. Boom!" And they did so because of her "[p]ublic depravity"—because of behavior that Melinda earlier in the story characterizes as "whorish." Although the drones may in ways operate as security devices to protect citizens of nations, Smith invites readers to question the meaning of security and security's relationship to terrorism. She invites readers to question who gets protected from whom or what given that, in practice, these drones hunt down lower-class individuals—those who lack privilege to own digital devices and those who lack digital literacies. These drones mark an incipient moment in what Smith characterizes as horrific emergent history—a moment that speaks to Incipio's name. Indeed, since these drones aim to annihilate the poor, the future toward which they strive is one in which wealthy, digitally connected citizens venerate digital technology and ignore unjust acts of violence.

Smith's readers also see the ways in which digital technology and globalization foster emotionally and physically violent behavior among local and global citizens. Most notably, the author demonstrates the effects of globalization on locals through her representation of Melinda's "ancient debt infraction" and the kind of emotional abuse that emerges out of her low social-class status. She abandons Aggie with Bill instead of accompanying her to Maud's funeral, indicating that the emotional abuse she experiences at the hands of a stratified society—abuse that forces her to struggle constantly to survive—begets related emotional abuse. Similarly, Bill as a global citizen behaves in unethical ways because digital games pervade his psyche. Evocative of the first-person shooter game that Maxine Tarnow's sons play in Pynchon's *Bleeding Edge*, Blood Head 4 as Smith represents it normalizes and makes child's play of apocalyptic violence for Bill. It contributes toward transforming Bill into a machine that can no longer empathize or experience emotion, and it primes him to develop into a real-life soldier such as Maud's former lover, Jimmy Kane, who, according to Aggie, had a device that resembles the one Bill has and used it for militaristic and recreational purposes. (He would make Maud "nicer to look at when they were doing it.") As Bill plays Blood Head 4, he engages in a precursor to militaristic violence that Jimmy engaged and perhaps still engages in. For instance, in the game, Bill places "a number of grenades about his person." And he gets into a firefight with Russian commandos when they come into view in a moment that invokes hostility between the United States and the Soviet Union during the Cold War, which gives birth to an arms race that produces technological developments such as ARPANET (the original internet). When he encounters the "traitor Vice-President," he takes him hostage and drags him "down the Mall with a knife to his neck." The game changes Bill physically in virtual reality—or at least it allows him to change himself. Because it allows him to give himself breasts, it enables him to push the limits of his gender in the way that the adolescent, male Sam Bloch overcomes them through his Latina avatar, Samanta, in Foer's *Here I Am*. It allows him to push the limits of his humanity as well when he shoots knives out of his wrists and gives himself a "scaled tail." This physical transformation to Bill's body in virtual reality speaks to Bill's real-life ideological transformation, which involves his unsettling disconnection from the real effects of violence. For instance, when Smith's readers learn from the narrator that Bill knows the sound that "a small animal makes when, out of sheer boredom, you break its leg," they learn that an act such as disemboweling a fawn in Blood Head 4—something that Bill does—can evidently lead to violence against animals or even people in real life.

Finally, Smith intimates that digital technology and globalization transform nature and perspectives on the natural world much as they transform humanity. Bill functions as a quintessentially digital citizen who prefers life in virtual reality as opposed to life in nature. According to the narrator, Bill dislikes "those things which crawled and slithered upon the earth." When he sees "tiny spirals on the sand," he indicates his aversion to them by comparing them to "miniature turds stretching out to the horizon." Perhaps because he prefers technology to nature, he comes to view technology as natural, as evidenced, for instance, by the narrator's descriptions of Bill's perspective on the drones. As the narrator explains early in the story, the drones dive "low like seabirds after a fish" while Bill stands watching them. Later, the narrator indicates that after the drones "were finishing their sallies"—strategic militaristic movements made historically by physically inhabited aircrafts—they "had clustered like bees." In turn, Smith shows how Bill views nature as technological, complementing these similes that involve the drones with descriptions of technology as literally mediating nature and likewise complementing Eggers's portrayal of the fusion of technology and nature in *The Circle*. Indeed, Bill continually examines the natural world through the lens of his digital device, which, for instance, identifies "*Arenicola marina*" for him—sandworms that are visible on the Felixstowe beach at low tide. He seeks digital information about natural phenomena more so than he seeks engagement with nature. Through virtual reality, he drowns out the natural world, which is decaying due to corporate disregard for environmentally friendly modes of manufacturing and distributing materials. According to Smith, "oil streaks" cover the sand on which Bill walks, but Bill has little regard for either the sand or the global problem that it showcases. Instead, with his AG 12, he covers the coastline before him with "a gleaming pavement, lined on either side by the National Guard, saluting him" as he makes his way toward the White House in the virtual reality of Blood Head 4. He replaces real nature with virtual manufactured material because his digital device allows him to live out what Smith sees as a capitalist American or globalizer's fantasy of geographical transformation to ecologically devastating ends.

Smith complements her portrayal of the potential for apocalypse as global devastation with a portrayal of the potential for apocalypse in another sense of the term: as a revelation, perhaps about the adverse effects of digital technology and globalization as the United States arguably put the world on course to witness them. Throughout the story, and evocative of considerations of maturity in Shteyngart's *Super Sad True Love Story* and Foer's *Here I Am*, the narrator fashions the narrative as a bildungsroman in which readers see how characters' looming futures render them as in medias res within the world of the story: in the process or on the verge of becoming something, although what they might become remains unclear. These characters are moving toward the possibility of a revelation as they come of age, much as globalizers might see the planet as coming of age in its state of interconnection with the contemporary moment. For instance, the narrator specifically draws attention to Aggie's impending age as opposed to stating her current one. She notes not that Aggie is eight but that she will be "nine in two days." Later in the story, Bill reinforces the narrator's sense of Aggie as being on the verge of a new age when he thinks of her as "an almost-nine-year-old." He reinforces the notion that new insights might accompany her birthday. Similarly, in mentioning Bill's age, the narrator observes that he "would be fifteen in May, almost a man!" Although the age of fifteen hardly constitutes adulthood, the narrator's remark conveys Bill's longing to leave childhood for a new developmental stage. The narrator then echoes this language and suggests more explicitly that a connection exists between aging into adulthood and having revelatory experiences, which schools such as Bill's apparently commodify as milestones to be met and which this school's teachers and students alike view as existing on a predictable timeline. As the narrator expresses it, Bill "was almost fifteen, almost a man, and the great human mysteries of this world were striking him with satisfying regularity, as was correct for his stage of development."

For Smith, virtual and material culture vie for control over the human imagination and the development of human history after the end of history (to reference Francis Fukuyama's term) and in the aftermath of the American Century.[11] This struggle between cultures shapes the process of revelation about the problems of the digital, globalized age. In Bill's experience in the story, this struggle begins to manifest itself when he is en route to the laying out in physical reality while he is at the door of the Oval Office in virtual reality. At this point, he opts to "split the visual" on his AG 12 "in order to pause and once more appreciate the human mysteries of this world slash how far he'd come." After his field of vision slashes into two—with virtual reality on one side and physical reality on the other—another kind of slashing also begins. When Bill arrives at the wake at St. Jude's, the site of a "local, outlier congregation" that is named after the patron saint of lost causes—the saint who would be best poised to help Bill as a lost cause—Smith as author works to slash into Bill's worldview through the physical reality via which she presents him. In St. Jude's, Bill finds himself in a dead zone in multiple senses of the term. First, he finds himself in a space that is, at the moment he arrives, dedicated to mourning the dead. Second, he finds himself disconnected from the digital network that allows his device to work. As he "stepped forward like a king" toward the president in the Oval Office to receive a salutation and handshake to signify successful completion of his digital game, "the light was failing, and then failed again; the celebrations were lost in infuriating darkness." As Smith continues, "The boy touched his temple, hot with rage," in a struggle to stay connected to the celebration of violence in virtual reality. Later, his digital device fails to seamlessly transmit a message from his father. As Bill touches "his sweaty temple" in a struggle "to focus on a long message from his father—something about a successful inspection and Mexico in the morning," Bill fails to fully process what his father attempts to convey about further transcending national divides physically by traveling the globe.

Smith suggests that human networks and human mysteries likewise shape revelation, and hence they can or should complement—or even supplant—the digital networks that make possible digitization and globalization as mysterious phenomena that define the twenty-first century and morph humans who have potential for humane actions into lost causes. In St. Jude's, Bill encounters human mysteries in the form of old objects in a well-worn space that has strategic overlap with the virtual American and postapocalyptic landscape that has heretofore absorbed his imagination in the story. He encounters a worn, putrid, local reality that elite members of society rarely have to see and that might not typically function as revelatory per se. As the narrator expresses it in describing what Bill sees, "a low-ceilinged parlor came into view, with its filthy window, further shaded by a ragged net curtain, the whole musty hovel lit by candles." Moreover, a network of decaying bodies fills the space as a metaphorical complement to the digital networks that Bill knows well. According to the narrator, Bill "couldn't even extend an arm—there were people everywhere, local, offensive to the nose, to all other senses." These bodies move him along a trajectory that he would rather avoid. In the narrator's words, Bill is "pushed by many hands, ever forward," much as history as Pynchon and Egan portray it has heretofore propelled humanity ever forward toward globalized and digitally interconnected realities—albeit without allowing citizens the opportunity to develop a sufficient philosophy about responsible modes of interconnectivity. He is pushed toward a moment of face-to-face connection with Maud's corpse, a lost cause of a different kind that rests in a well-worn coffin, which suits her as an impoverished victim of global capitalism's exploitations. Her body as a material thing is beyond physical salvation as it lies in "a long box, made of the kind of wood you saw washed up on the beach."

Smith implies that Bill and global citizens who resemble him might not be lost causes if they treat their encounters with physical reality as revelatory and, more to the point, if they come to terms with death's reality in the face of digital technology that virtualizes it, desensitizes them to it, and causes it. Bill certainly knows something about death from Blood Head 4, where "many bodies were lying on the ground" of the Oval Office and elsewhere in postapocalyptic Washington, DC. Bill likewise knows something about death in life by virtue of the fact that his mother is dead. Even though he never knew her or saw her dead body per se, he knows the experience of growing up motherless. It contributes toward shaping him as a self-centered and callous adolescent. Yet his encounter with Maud's dead body at the wake only allows Bill to experience a limited revelation. He certainly develops a clearer sense of death's inevitability when he sees it all "clearly in the candlelight—the people in black, weeping, and Aggie on her knees by the table, and inside the driftwood box the lifeless body of a real girl." As the narrator continues, the body was "the first object of its kind that young Bill Peek had ever seen," drawing attention to the dramatic difference between the numerous virtual deaths that Bill encounters and the physical reality of a dead body in life. But he fails to see the connection between virtual and physical death. He fails to see that digital technology which enables the existence of virtual reality also enables state-sanctioned violence against exploited citizens within national bounds that results in the killing of real, innocent locals such as Maud.

Smith likewise indicates that Bill and global citizens who resemble him in that they move between virtual and physical reality might transcend their status as lost causes if they experience a spiritual revelation and embrace spiritual reality as a complement to physical and virtual reality. Notably, the narrator's description of the song the crowd sings while Bill moves through the space draws attention to the problems of realizing a spiritual life—problems that come to a head in a globalized world that fuels nationalistic and xenophobic impulses. They sing *"Because I do not hope to turn again . . . Because I do not hope,"* the opening lines from T. S. Eliot's "Ash-Wednesday," a poem that evokes the holiday that marks the first day of Lent and the beginning of the passion of Christ ("Meet the President!"). The poem, too, addresses the subject of religious conversion because Eliot wrote it following his own conversion to Anglicanism in 1927. Smith's interest in the poem likely stems from the fact that the speaker of Eliot's poem experiences what John Kwan-Terry calls a "journey to the Absolute," which in some ways resembles Bill's journey toward Maud's body. Whereas the speaker of Eliot's text has a religious experience of conversion before the poem's beginning and struggles in the poem with "ambiguity and uncertainty" and the "lure of the world," Bill, in Smith's story, moves toward Maud's body in a process that might culminate in revelation and in a conversion of a sort—a conversion away from faith in the postapocalyptic conditions that the United States helps create and a conversion away from worshiping technology as a false god (Kwan-Terry 134). In other words, the revelation might culminate in a conversion toward a more spiritual and meaningful life. Indeed, the narrator's description of Maud's body from Bill's perspective suggests that moving from life into death involves spiritual transcendence in that it provides access to divine knowledge. In the narrator's words, a "slight smile" on Maud's face "revealed the gaps in her teeth, and suggested secret knowledge," which is knowledge that runs counter to the ubiquitous digital information that Ferris, DeLillo, and Eggers, for instance, critique and which is perhaps knowledge of the afterlife or God ("Meet the President!").

However, revelation again emerges as elusive in Smith's fictionalized dystopian future. Spiritual reality fails to complement physical and virtual reality in a sustainable way because digital technology and capitalist impulses pervade the consciousness of global citizens. Bill's experience of seeing Maud's smile mirrors that of the poet in Eliot's poem, who, according to Kwan-Terry, "is left, finally, in 'the time of tension between dying and birth'—the temporal sphere, the twilight zone of sequential time where there is no comprehensive understanding of anything," a sphere that is perhaps analogous to the virtual America of Blood Head 4 (138). Although Bill may see that a spiritual reality can and should complement physical and virtual reality, he fails to see how to develop a genuinely spiritual way of thinking and being for himself. He only shows a capacity for understanding spiritual transcendence through material, corporate terms that involve power (and money as power) above all else, terms that the United States so successfully established during the American Century. In describing the smile on Maud's face from Bill's perspective, the narrator notes that it bears a striking similarity to one that Bill "had seen before on the successful sons of powerful men with full clearance—the boys who never lose" ("Meet the President!"). According to the narrator, Bill feels "the sensation that there was someone or something else in that grim room, both unseen and present." But he sees no way in which the unseen presence might offer him a sense of fulfillment. He sees no way in which spirituality in a general sense or faith in God specifically might allow him to become the kind of adult he seeks to become within the world of Smith's apparent bildungsroman, which never fully realizes itself as a manifestation of the genre because Bill never matures within its pages as, for instance, Jacob Bloch finally matures by the conclusion of Foer's *Here I Am*. Instead of maturing to embrace spiritual knowledge, Bill regresses into fear of physical death. He sees the unseen presence only as "coming for him as much as for anybody." He comprehends it as a force that nullifies the privilege that his social class affords him, and he treats it as he treats the apparent villains he encounters in virtual reality.

Finally, Smith posits that a revelation of the value of human community might enable global citizens who lack a capacity for spirituality to find redemption. Although Smith predominantly shows interest in portraying members of a geographic community who lack a sense of interpersonal community in a work such as *NW*,[12] she draws attention, in "Meet the President!", to the ways in which communities comprised of individuals who have meaningful connections with one another have the capacity to perform meaningful and spiritual work in the world. For Smith, communities of people, more so than divine forces, create spiritual spaces and spirituality itself, and this notion most clearly comes to the fore of her story in the narrator's description of St. Jude's. By searching his AG 12 device, Bill comes to the conclusion that the site of St. Jude's is "[n]ot a church" because it is located in a building that was "originally domestic property, situated on a floodplain, condemned for safety." The so-called church has "no official status" according to Bill's device. But Smith's readers see what Bill cannot when he arrives at the space: that it *is* a church because locals converge in it with spiritual purpose. The space unites them as nationalism and globalization cannot. They come together by caring for Aggie, approaching her to inquire about her welfare and paying respect to her sister in a way that, to echo the words of one local, "[d]oes the soul good." Thus, much as *NW* finally "succeeds in imagining community [. . .] in the event of Colin Hanwell's funeral," "Meet the President!" succeeds in showcasing community at this funeral of another fictionalized Hanwell (Kruger 77). Through the scene, Smith's readers see that a church, like knowledge in Ferris's *To Rise Again at a Decent Hour*, exists as a social construct. It exists as such because human communities have power to render the mundane and the human as sacred. Furthermore,

Smith's readers see potential in well-worn communal spaces that juxtapose with the "clean, blank places" that Bill says he prefers early in the story—spaces where Bill as a user of digital technology is "free to fully extend, unhindered" by national divides or anything at all, for that matter, but spaces that fail to allow him to mature emotionally or spiritually.

Ultimately, Smith invites her readers to achieve a real future that her postapocalyptic, dystopian, fictionalized future presents as a relative impossibility. Her readers, like Bill, encounter the contemplative verses of Eliot's "Ash-Wednesday"; they, like Bill, encounter Maud's dead body, albeit in the form of a literary representation; and they, like Bill, encounter the compelling vision of local human community at St. Jude's as they interact with Smith's old-media print text—or perhaps an online version of it on the *New Yorker*'s website. Hence readers might treat Smith's literary text as a revelation by genre and develop responses to it that signify their having experienced revelation as a moment of deeper seeing and understanding. These responses might involve engagement with real as opposed to digital life. Too, they might involve the cultivation of real community, a process that begins with rejection of the culture of being alone together. Although the digital, globalized age has, in Turkle's words, allowed users of digital technology to "hide from each other" through "networked life" behind figurative screens that function as borders, Smith seems to hope that her readers can envision something beyond the limits of their attraction to the apocalypse and mass communal solitude as a form of apocalypse (*Alone Together* 1).[13] As Smith articulates it in "Generation Why?" "We were going to live online. It was going to be extraordinary. Yet what kind of living is this? Step back from your Facebook Wall for a moment: Doesn't it, suddenly, look a little ridiculous? *Your* life in *this* format?" A better format beyond a wall or screen, as Smith sees it, is precisely what the world needs. And, as she suggests, her readers can realize this better and metaphorically borderless format through socially responsible, nondigital, and local ways of thinking and living—ways of being that reclaim existence from the clutches of the digital and redefine what it means to make personal and global connections and social progress.

Digital Screens and National Divides in Mohsin Hamid's *Exit West*

Mohsin Hamid's *Exit West* (2017), the focus of the second part of this chapter, resembles Smith's story in that it portrays the problem of national divides and the ways in which screen culture simultaneously diminishes and reifies those divides. Hamid begins his novel with the story of Saeed and Nadia's first encounter in a business course that speaks to Hamid's subject of globalization.[14] It then proceeds to depict the development of Saeed and Nadia's relationship as they migrate across a paradoxically globalized yet bordered world that is analogous to the world of Smith's story—a world in which, like Bill, they are tethered to digital devices that shape *Exit West* as a work of magical realism. Much as the internet "really is magic" for characters in Pynchon's *Bleeding Edge*, digital devices in Hamid's text, which paradoxically aren't fantastical but real, help Hamid develop a fantastical element in his work (*Bleeding Edge* 398). These devices show that what was once a digital divide—a difference in the access that individuals had to digital technology—diminishes as a result of twenty-first-century technology's ubiquity. Indeed, much as they do in DeLillo's *Cosmopolis*, digital devices allow users across social classes in Hamid's novel to experience transcendence, especially amid the otherwise technologically stunted postindependence circumstances of the unnamed city of Hamid's text, where despite the pervasiveness of digital technology, a hardwired telephone line "remained a rare thing" (39). These devices transport a city perhaps akin to Lahore as Hamid represents it in *Moth Smoke* and *The Reluctant Fundamentalist* or the unnamed city of *How to Get Filthy Rich in Rising Asia* into a more equitable future beyond a backlash to globalization that involves violence and corruption. As Hamid expresses it, "Nadia and Saeed were, back then, always in possession of their phones," which resemble "wands" waving "in the city's air" (39). Their phones' "antennas sniffed out an invisible world, as if by magic, a world that was all around them, and also nowhere, transporting them to places distant and near, and to places that had never been and would never be" (39).

Yet Hamid complements his romanticization of digital technology as a magical phenomenon with criticism of it that speaks to Smith's judgments and the forms of critiques that appear as conventions of magical realism as a genre.[15] He criticizes it for the ways in which it alters perception and is addictive, intimating that a link exists between the effects of digital devices and drugs. Turkle illuminates this connection, observing that when "children were introduced to video games in the 1980s, there was serious discussion of banning them using the same statutes that outlawed addictive substances such as heroin and marijuana" (*The Second Self* 4). For Hamid, a connection between apparently magical digital technology and drug use appears to manifest in the "feeling of awe" that both create (*Exit West* 46). Saeed develops this feeling in Hamid's novel upon consuming magic mushrooms, which Nadia purchases for herself and Saeed early in their relationship. When Saeed takes the magic mushrooms that Nadia buys, he feels a sense of "wonder" about features of the world around him that he otherwise overlooks (46). And he disconnects from reality much as a user of digital technology does. He resembles Maxine Tarnow, the protagonist of Pynchon's *Bleeding Edge*, who, after getting lost in the fictionalized internet subspace known as DeepArcher, exclaims "Holy shit. What time is it?" (77). He loses a sense of the real world around him in the way that an internet user enters into what Bauman terms "the 'great simplification'": the worry-free state that an online world enables in the age of mass migration, a state in which "one is saved from the inevitability of confronting the adversary point blank" (Bauman 106). He forgets his phone's existence, his nation's problems, and his parents' concerns. When he recharges his phone, it chirps "with his parents' panic, their missed calls, their messages, their mounting terror at a child not returned safely that night, a night when many children of many parents did not return at all" due to violent clashes in the unnamed city (Hamid, *Exit West* 47–48).

Hamid showcases different approaches to negotiating the addictive nature of digital technology in his fiction with the goal of underscoring that human relationships with technology can and should involve negotiation so that technology's development never outpaces the development of philosophy regarding its use. Notably, characters in his novel perceive and use their cell and smartphones in distinctively different ways, much as Jacob and Tamir in Foer's *Here I Am* do. Hamid's characters do so in large part because they see that their phones function to enchant them—for better or for worse. They see what Turkle characterizes as technology's ability to change "people's awareness of themselves, of one another, of their relationship with the world" (*The Second Self* 18–19). Whereas Nadia sees "no need to limit" her use of her phone, riding it "far out into the world on otherwise solitary, stationary nights," and thereby celebrating it unabashedly for the access it provides, Saeed in part resists "the pull of his phone" because he questions its virtues (*Exit West* 41, 39). The narrator explains that Saeed "found the antenna too powerful, the magic it summoned too mesmerizing, as though he were eating a banquet of limitless food, stuffing himself, stuffing himself, until he felt dazed and sick, and so he had removed or hidden or restricted all but a few applications" (39–40). He sees what Bill in Smith's "Meet the President!" or Mae Holland, the protagonist of Eggers's *The Circle*, never manage to see in their respective experiences with the digital world: that digital dystopia exists as a possibility because digital devices allow their users to become "present without presence," in Hamid's words (40).

Hamid's concept of being present without presence speaks to Bauman's critical remark in *Strangers at Our Door* that "[q]uite often we manage to be in" two "worlds at the time," for instance when we are "sitting at the family table or walking in the street, alone or in a group, while exchanging tweets with a Facebook friend hundreds of kilometres away" (103). And it resembles Turkle's sense of twenty-first-century citizens as being alone together—as hiding "from each other" by way of "networked life" (*Alone Together* 1). Like Bauman and Turkle, Hamid characterizes this problem of being present without presence as one that humanity must address to avoid a dystopian reality. Certainly, being present without presence sustains an allure as Hamid represents it in that it allows those who live apart from one another to traverse the limits of geography. For instance, Saeed and Nadia stay in touch in the unnamed city despite religious restrictions on contact between men and women. Technology allows them a certain degree of freedom to communicate and to defy oppressive laws that in earlier moments in history may have functioned to snuff out their personal connection with each other. Digital technology also allows users to traverse geographical distance, as evidenced by Saeed and Nadia's experience of migration. Hamid repeatedly presents Saeed as attempting to check in on his father by cell phone after his move away from the unnamed city. He depicts him as capable, in theory, of sustaining a tie that might utterly dissipate without digital technology. However, being present without presence through digital devices allows users to disconnect at will and to problematic ends. In Turkle's terms, this "virtual intimacy" may "degrade our experience of the other kind and, indeed, of all encounters, of any kind" (*Alone Together* 12). Moreover, digital technology users' *choice* to be present without presence through digital devices is peculiar because forced migrants often lack such a choice. When forced migrants communicate with support networks using Facebook groups and WhatsApp, they do so with the hope of eventually attaining documentation to legitimize their full presence and hence their existence in xenophobic nations that predominantly seek to keep them out (Kingsley 238).

The paradox of being present without presence haunts Hamid's representation of state-sanctioned and terrorist violence that involves technology—violence that renders psychological and emotional disconnection as necessary for survival. In Hamid's text, cell phones appear as weapons when Saeed and Nadia rest their phones "screens-down between them, like the weapons of desperadoes at a parley" (17). And this portrayal of phones primes readers to notice the weaponization of digital technology in *Exit West*. For instance, "media-savvy" militants take over the unnamed city's stock exchange and create a digital-media "spectacle" before the building is "stormed with maximum force," an event for which television viewers are present without presence (43). Likewise, the government creates widespread anxiety by turning off mobile phone signals and internet access as a "temporary antiterrorism measure" in the unnamed city, rendering those who rely on phones as present in their unnamed city but not present in their typical channels of communication (57). The government reifies a digital divide that has increasingly dissipated according to Hamid's portrayal. And "flying robots"—drones "high above in the darkening sky" akin to those in Smith's "Meet the President!"—engage in surveillance and threaten sudden attack (88). They render the national military as near-ubiquitously present through technology while not physically present. These apparently typical occurrences leave citizens of the unnamed city such as Saeed and

Nadia to feel "marooned and alone and much more afraid" (57). They also lead them to begin the process of imagining an escape into the West, a world that is ostensibly more free of the problems of war and terror. They live their physical lives as circumscribed by the violence of their Eastern home while their imaginations disconnect, struggling westward toward a dream of emotional and psychological stability.

But the West as Hamid represents it never functions as the promised land that citizens of the East imagine it to be because of the interplay of problems involving technology, violence, and forced migration. In the West, technology and violence sustain a connection to each other and produce an effect of being present without presence that in ways mirrors the effect that exists in the East, suggesting that Hamid views the East and the West as similar and not as existing in the kind of dualistic opposition that twenty-first-century mass media aim to propagate and exploit. Saeed and Nadia emerge as analogous to historical migrants with whom Hamid would be familiar through his experiences in Pakistan, a nation that bore witness to forced migration upon the "creation of India and Pakistan," which "led to the flight or expulsion of 8 million people" (Hansen 256). Pakistan also bore witness to forced migration upon the United States' invasion of Afghanistan as part of the War on Terror, which led to a protracted refugee situation with Afghans fleeing to Pakistan (Betts 77). In their own fictionalized experience of forced migration, Saeed and Nadia simply come to trade one alienating and violent landscape for another when they enter nations as refugees who are present physically without having legal presence. In London in particular, violence evocative of Saeed and Nadia's home city manifests as "soldiers and armored vehicles" invade the refugees' neighborhood (*Exit West* 137). And "drones and helicopters" threaten violence from on high much as drones did in the East (137). A mob of natives even echoes the actions of militants from Saeed and Nadia's Eastern home. To Nadia, this mob looks "like a strange and violent tribe, intent on their destruction, some armed with iron bars or knives" (134). This machinery of war and persistent hostility by natives shape conditions for migrants to dream of exiting further west under London's "drone-crossed sky": to the United States (188). It leaves them in circumstances similar to those that prompted them to exit west in the first place.

The apparently progressive United States, too, fails to function as a promised land for Hamid's fictionalized migrants—even though the concept of what counts as a native in the United States is arguably more complicated than in some other nations. Hamid states that many Native Americans "died out" or were "exterminated long ago," but a nativist or racist mind-set persists in the nation of migrants, and hence the absence of agency for migrants persists as well (197). This racist mind-set alienates migrants to the United States who cannot escape the condition of being present without presence. Although no mobs attack Saeed or Nadia, Nadia encounters a "pale-skinned tattooed man" who enters the co-op where she works and places his "pistol on the counter" in an act of intimidation and aggression (215, 216). Saeed and Nadia likewise live life under digital surveillance in the San Francisco Bay Area much as they lived under surveillance in the unnamed city and in London and much as Eggers portrays Bay Area Americans as living under the Circle's surveillance in *The Circle*. When "one of the tiny drones that kept a watch on their district, part of a swarm," crashes into the "transparent plastic flap" of a door to their shanty home, they opt to bury it as though it is a human soldier "in the hilly soil where it had fallen" (205). They perhaps put to rest not only the humanized bits of digital-age war machinery, but the notion that escape from a vast if not ubiquitous network of violence exists as a possibility in the twenty-first century.

Hamid stretches the concept of connectivity beyond digital parameters by way of his depiction of the connections that migrants retain to their homes as they migrate westward into racist and xenophobic nations. Most notably, Saeed's links to the unnamed city and the cultural and religious features that define it frequently leave him disconnected from circumstances in London and Marin. Unlike Nadia, who never really practices the religion of her home and who seems to feel a sense of liberation in leaving the gender-based oppression that defines her experience in the predominantly Muslim unnamed city, as evidenced by the narrator's remark that "she had been stifled in the place of her birth for virtually her entire life," Saeed tends to long for the cultural and religious community that he had prior to migration (159). He longs for religious community that resembles the kind of community Ferris portrays the protagonist of *To Rise Again at a Decent Hour* as longing for. In London, he tells Nadia that he wants to "be among our own kind," by which he means people from "the country we used to be from" (153). He even finds a house of his fellow countrymen to which he wants to move even though Nadia views the house as occupied by "dozens of strangers" and even though these men remind him at points of the "militants back home" (153, 155). He longs for prayer among his own people, and as the narrator describes it, praying among them feels different for Saeed in London because it makes him "feel part of something, not just something spiritual, but something human, part of this group" (152). In Marin, too, he engages in prayer as a means by which to connect to his past, of which his culture and nation are an inevitable part. As the narrator remarks, "he prayed fundamentally as a gesture of love for what had gone and would go and could be loved in no other way. When he prayed he touched his [now-dead] parents, who could not otherwise be touched, and he touched a feeling that we are all children who lose our parents, all of us, every man and woman and boy and girl" (202).

In turn, Hamid proposes that migrants who retain connections to home experience nostalgia, a metaphorical manifestation of being present without presence in that it inhibits their engagement in present-day human relationships, which Hamid eventually comes to represent in digital terms. The more Saeed clings to his past and to the culture and faith that shape his sense of that past, the less he connects with Nadia, for instance when they arrive in Greece and experience the landscape simultaneously alone and together, to again reference Turkle's book's title. As Hamid depicts it, "In the late afternoon, Saeed went to the top of the hill, and Nadia went to the top of the hill, and there they gazed out over the island, and out to sea, and he stood beside where she stood, and she stood beside where he stood" (108). But as they "looked around at each other," they "did not see each other, for she went up before him, and he went up after her, and they were each at the crest of the hill only briefly, and at different times" (109). By the time they reach London, they come to speak to one another with "unkindness" and lose their romantic connection (132). As they discuss the possibility of moving from the diverse house of refugees in London to the house occupied exclusively by natives of Saeed and Nadia's home country, Saeed opts against bridging their disconnect—"the tiny distance it would have taken to kiss" Nadia (153). And it is the very process of migration that leads them and migrants who resemble them to see one another differently. Hamid observes that "[e]very time a couple moves they begin, if their attention is still drawn to one another, to see each other differently, for personalities are not a single immutable color, like white or blue, but rather illuminated screens, and the shades we reflect depend much on what is around us" (186). Indeed, divisive screens form between individuals who in Hamid's description resemble screens of digital devices that, in addition to providing access to the digital world, also reflect their material surroundings. Once Saeed and Nadia arrive in the Bay Area, Nadia "was herself putting barriers" between her and Saeed (196). She screens him out.

The numerous screens, barriers, and disconnections that appear consistently throughout *Exit West* indicate Hamid's interest in showcasing a duality or dividedness that shapes personal psychologies in the digital age. Disconnection from homelands and attempts at making connections in new nations lead Saeed to exhibit a disconnect between what he wants or thinks he wants to do and what he actually does—the type of disconnect that Kristen Roupenian portrays in "Cat Person." For instance, when a tough Nigerian woman in Marin in a community of refugees that Nadia joins blocks Saeed's path in an emasculating act that upends Saeed's sense of traditional gender roles as his home propagates them, he sits on his bed with his heart racing and wants "to shout and to huddle in a corner but of course he [does] neither" (151). He appears divided among the traditionalism of his past, the relative progressivism he witnesses in his present, and a future that he inevitably still views as uncertain. Disconnection exists within him as much as it exists between him and the nation he inhabited as well as the nation he presently inhabits. Nadia experiences a sense of dividedness akin to Saeed's while reading news of the power outage on her smartphone in London, a disruption to electrical service to refugees that allows the government to flex its metaphorical muscles and a symbol of the relationship between power in the form of electricity and power in the form of influence. Hamid makes a point of underscoring the role of digital media in shaping Nadia's divided state. As Nadia sits on a stoop with her phone in hand, she has an experience that is evocative of "The Harmony" in Foer's *Here I Am*—the moment when natural light in virtual reality and physical reality correspond (*Here I Am* 62). She thinks she sees "online a photograph of herself sitting on the steps of a building reading the news on her phone across the street from a detachment of troops and a tank, and she was startled, and wondered how this could be, how she could both read this news and be this news, and how the newspaper could have published this image of her instantaneously" (*Exit West* 157). As Nadia examines her surroundings in search of a photographer or perhaps even a drone that may have taken the photograph, she has "the bizarre feeling of time bending all around her, as though she was from the past reading about the future, or from the future reading about the past, and she almost felt that if she got up and walked home at this moment there would be two Nadias" and that "two different lives would unfold for these two different selves" (157). Although she comes to realize that the photograph she is viewing is of another woman and not her, Hamid's readers see that her reality greatly resembles the other woman's. And they also see ways in which the past, present, and future emerge as confounded due to the interplay of migration and digital technology. Hence they develop a sense of the uncertainty that twenty-first-century individuals' divided experiences perhaps inevitably involve.

Hamid complements his portrayals of individual experiences of dividedness with a representation of dividedness as it comes to exist on national and global scales and as it involves digital connectivity and race. Most notably, in describing the orchestrated power outage in London, Hamid draws attention to the way in which British government officials divided London into dark and light communities in more ways than one. The outage retools the old binary between East and West into one that divides migrants who are predominantly people of color from natives who are predominantly white and racist. It also retools the digital divide: if have-nots can have digital devices in the twenty-first century as Hamid represents it, the government will rob them of electricity that allows them to power those devices. Hamid shows these government officials as positioning the predominantly white natives who live in "light London" as the sanctioned and empowered Londoners; in turn, the officials position the migrant, presumably predominantly nonwhite residents of "dark London" as disenfranchised and unwelcome inhabitants in a peripheral and disposable part of the city (146). This division of London speaks to the social class divisions in Felixstowe that Smith represents in "Meet the President!" It likewise showcases the ways in which England on the whole becomes divided and "like a person with multiple personalities" as a result of migrants' presence in the nation, with some citizens "insisting on union and some on disintegration" (158). As Hamid continues, "in fact, some said Britain had already split, like a man whose head had been chopped off and yet still stood" (158). England's divided state thereby emerges as indicative of a dividedness that may, too, shape the fates of other nations because, as Hamid articulates, "as everyone was coming together" in the digital age of migration, "everyone was also moving apart" (158). As globalization works apparently to connect disparate lands and people, those people come into conflict with one another and their differences starkly divide them, especially with regard to views on what it means to be a refugee and on solutions to the migration crisis.

Whereas Smith shows interest in spirituality as a potential solution to the problems of digitization in "Meet the President!", Hamid shows interest in the aesthetic features of art and of novels in particular much as Shteyngart in *Super Sad True Love Story*, DeLillo in *Cosmopolis*, and Eggers in *The Circle* show interest in them. He sees them as a means by which to imagine possibilities for meaningful connectivity that can supplant the disconnectedness which defines Saeed and Nadia's experiences of migration in a supposedly globalized and interconnected world, but a world that, according to Bauman, allows "[l]oners in front of a phone, tablet, or laptop screen" to "put reason together with morality to sleep" (108). Among the most noteworthy of these aesthetic features is Hamid's book's form as a novel in the digital age, an age in which, to quote Jonathan Franzen's apocalyptic proclamation in "The Reader in Exile," "a viewer is born" for "every reader who dies" (165). As a novel that focuses heavily on digital culture and connectivity in content, Hamid's book functions at first glance as somewhat of a paradox. A novel might be seen as a peculiar vehicle for what I propose is a nuanced literary argument about connectivity and screens of different kinds, especially given that Hamid has ventured into digital publishing that fits with standard definitions of digital humanities via "The (Former) General in His Labyrinth," an interactive digital story that Penguin released in 2008 as part of its digital fiction project. Readers who read Hamid's digital tale during its time online could contemplate thematic connections between Hamid's narrative and *Tales from the Thousand and One Nights*, a collection of folktales from the Middle Eastern and South Asian Muslim world narrated by the fictionalized Scheherazade. Hamid's readers, too, could contemplate digital connectivity as a form as they clicked their way through a digital path around a visual representation of the fictionalized General's palace. But like most digital objects that continually morph or disappear as web editors make adjustments to layouts, images, texts, and web addresses, "The (Former) General in His Labyrinth" and the arguments it makes about connectivity disappeared, except perhaps for internet-savvy web users who can navigate to it via the Wayback Machine. Unless readers have a good memory and can turn a digital tale into a contemporary folktale, their contemplation of the text can only last as long as the digital world fosters the text's existence. Thus, a notable virtue of *Exit West*'s aesthetic form as a novel, especially as a *print* novel, is its endurance. By way of its physical form, it makes a more enduring connection with readers because it lives in the material world and not just the digital one, eluding what Pynchon might call the "Deep Web"—the shrouded underworld of the internet that now houses and in large part conceals from mass view "The (Former) General in His Labyrinth" (*Bleeding Edge* 10). It speaks to what Alexander Starre's *Metamedia* suggests is "a resurgence of the book" amid the digital age, but without the visual design elements that most interest Starre (27).

Moreover, as a novel, *Exit West* employs aesthetic features that comment on connectivity as a subject: vignettes that periodically interrupt the main narrative about Saeed and Nadia while inviting readers to contemplate the short attention spans that the digital age has fostered—the expectation of or even desire for interruption—and, more notably, the connections and disconnections that accompany the interruptions they create. For example, in transitioning into the first of these vignettes, Hamid draws attention to the contemporaneity of apparently disparate events and relationships between them. He likewise draws attention to the different kinds of networks that define the times as connected—networks that power digital-age technology as well as those that connect apparently disconnected nations and people. Hamid writes that as "Saeed's email was being downloaded from a server and read by his client," a "pale-skinned woman" is sleeping in a room "bathed in the glow of her computer charger and wireless router."—a room in her home in Surry Hills in Sydney, Australia, a recently gentrified neighborhood that allows her to disarm the alarm system that previous occupants had installed and likely left armed (*Exit West* 7, 7, 8). The narrative reads as simultaneously interrupted but connected by Hamid's transition. And the email exchanges, chargers, routers, and alarm systems of the scene invite readers to contemplate various technological connections. They enable Hamid to complement the thematic content about connectivity that appears in the main narrative with similar thematic content in the vignette. This content, too, invites readers to draw connections between circumstances that prompt migration out of war-torn nations and circumstances that enable comfortable existence for white citizens in gentrified neighborhoods such as Surry Hills. As the pale-skinned woman sleeps safe and comfortable in her home and in her relatively privileged circumstances, a dark-skinned man with "woolly hair" emerges out of a magical door into her closet, perhaps leaving home due to violence, racism, xenophobia, or some combination thereof (8). Hence Hamid leads readers to draw connections between this vignette and other thematically similar moments in his novel, in the main narrative or in other vignettes. The disruption or disconnection leads to connection as readers may consider, for example, the discussion of dark and light London in the story of Saeed and Nadia; the vignette about a family with "dark skin" that migrates to Dubai and gets "lost in an aura of whiteness" on the security camera screen that monitors its movements (91); or the vignette about the whiskey drinker in the Shinjuku district in Tokyo who "disliked Filipinos" and may well be planning to enact racially motivated violence against Filipino girls who just migrated through a magical door (31).

Furthermore, many of Hamid's vignettes appear as aesthetically unfinished or ambiguous, and the incomplete state in which he presents these vignettes speaks to uncertainty as it comes to define circumstances for twenty-first-century global citizens in the digital age. Indeed, the Tokyo vignette ends in medias res with the whiskey drinker following the Filipino girls, slipping "into a walk behind them, fingering the metal in his pocket as he went" (31). Hamid's readers never learn whether or not the Filipino girls survive their migration beyond their apparent first moments in Tokyo because the vignette ends in medias res. They never learn whether the whiskey drinker attacks or murders them. This uncertainty and ambiguity exists as anomalous in a digital age in which apparently complete stories, answers, and information are ever readily available online even though their veracity remains elusive, a reality toward which Ferris gestures in *To Rise Again at a Decent Hour*. Similarly, unanswered questions remain for readers of the vignette about the old, presumably white, rich woman who lives in Palo Alto, California—and for the old woman as well. Hamid describes the old woman as living in her Palo Alto home her whole life, even though her children push her to sell her home because they want her money. He includes details about the way in which the old woman's favorite granddaughter "looked to the old woman, overall, more or less, but mostly more, Chinese" (208). Yet Hamid never explains whether a mixed-race marriage at some point in the granddaughter's history may have led to the granddaughter's Asian appearance. He portrays the woman as uncertain about her granddaughter's race, and he leaves readers uncertain about it as well—and perhaps contemplating the degree to which racial identity matters or *should* matter in the United States as a supposedly progressive Western nation. As a third example, the vignette Hamid includes about the maid from just outside of Marrakesh concludes without a definitive sense of whether or not she will ever leave to join her migrant daughter in Europe. Although readers assume that she will opt against joining her daughter because she has turned down many of her daughter's requests to do so, Hamid concludes the vignette and the chapter of which it is a part with the narrator observing that "[o]ne day she might go" but "not today" (226).

Ambiguous sentences provide an aesthetic analogue to Hamid's vignettes in that they, too, create a sense of uncertainty and also liminality, which scholars such as Bhabha in *The Location of Culture* have celebrated and authors such as Salman Rushdie in *The Satanic Verses* and in *East, West* have narrated. Hamid's own life experience leads him to shift existing discussions of hybridity to involve digital culture and hence to see twenty-first-century people as defined by their movements between material and virtual reality *and* between national borders that come to define individual and group identities. Key phrases in Hamid's text showcase this kind of movement and in-betweenness. For example, in the opening sentence of the novel, Hamid observes that in "a city swollen by refugees but still mostly at peace, or at least not yet openly at war, a young man met a young woman in a classroom and did not speak to her" (3). The narrator's phrasing suggests that the line between war and peace is blurred. And his phrasing echoes a style he employs at numerous later points in the novel, for instance when he observes that the eyes of a man coming through a door in a vignette "rolled terribly. Yes: terribly. Or perhaps not so terribly" (9). The narrator purposefully hedges. He inserts ambiguity into the narration in mentioning what otherwise exists as a minor detail of the story, and it is the ambiguity more so than the minor detail that captures the reader's attention, much as it does when the narrator notes, to provide a third example of this kind of aesthetic feature, that the displaced "were ashamed" but "did not yet know that shame, for the displaced, was a common feeling, and that there was, therefore, no particular shame in being ashamed" (184). The circuitous logic of Hamid's sentence leaves readers not only turning over his idea in their imaginations but turning over the reason for Hamid inviting them to do so. Readers realize that regardless of whether displaced persons do or do not feel shame, there exists something liminal, circuitous, and ambiguous about their condition and about that of twenty-first-century humanity amid pervasive digital culture in general perhaps because, as Clark suggests, physical humans are transformed by their own technological dependencies.

As aesthetic digital art objects that appear through ekphrasis (or verbal description) in the body of Hamid's novel, French photographer Thierry Cohen's photographs echo the uncertainties and in-between elements of Hamid's novel as a work of art; they counter representations of digital technology as mass consumed and detrimental; and they point to possibilities that uncertainty and ambiguity may afford in an age of screens and barriers of different kinds. Although Egan suggests in *A Visit from the Goon Squad* that "*digitization*" is "the problem"—that it creates a crisis in the twenty-first century because, among other things, it flattens the sound of music as a form of art—Hamid sees complexity in the possibilities that digitization presents, as evidenced by his portrayal of Cohen's digital photography (Egan, *A Visit* 23). He intimates that art about the digital age and digital art alike might provide new perspectives for their audiences. In the novel, Saeed shows Nadia *Villes éteintes* on his cell phone screen—digital photographs of "famous cities at night, lit only by the glow of the stars" (Hamid, *Exit West* 56). Cohen achieves the effect of natural lighting at night by removing unnatural light from the cityscapes using a computer. He then superimposes a night sky and lighting from a night sky from a place of the same latitude. The cityscapes that result are evocative of the preglobalized, predigital, and even premodern world, and they present a sense of uncertainty with regard to time and place. In other words, the cities in his photographs exist as independent of time in some way, or perhaps they show a postapocalyptic world because the beauty of nature as the night sky represents it manifests amid the physical evidence of modernity. Cohen's photographs show a noteworthy attention to perspective on a global scale, and viewers of his photography might consider possible arguments that his photographs make about perspective as a subject. Moreover, they show that there exists the possibility for beauty to develop amid confounding or confusing circumstances—or even because of these circumstances—because ambiguity affords viewers with new perspective. As Nadia thinks, the photographs "were achingly beautiful, these ghostly cities—New York, Rio, Shanghai, Paris—under their stains of stars, images as though from an epoch before electricity, but with the buildings of today. Whether they looked like the past, or the present, or the future, she couldn't decide" (57).

Hamid's novel echoes the aesthetics of Cohen's photographs in that it attempts to offer new perspectives, most notably perspectives that counter moral blindness, which Bauman sees as part and parcel of the migration crisis.[16] It portrays magic and beauty in acts of migration that predominantly appear as controversial in the media or as flattened by the digital screen, and it depicts migrants not as illegitimate or illegal but as part and parcel of that magic and beauty. More specifically, it portrays meaningful love, which Shteyngart celebrates in *Super Sad True Love Story*, as part of a vision for social justice in a twenty-first century defined by migration amid ubiquitous screens and barriers. In both the main narrative of Saeed and Nadia and in a vignette that appears near the end of *Exit West*, Hamid exhibits forms of love that counter the narrow limits of tradition and conservative perspectives (including the conservative Islamic perspectives that Hamid readily encounters) on what counts as a beautiful union. Although Nadia and Saeed disconnect from one another emotionally and drift apart much as the lovers of Shteyngart's novel do, through their migrations, they each find love and discover new perspectives on ways in which they can love. In California, despite "some resistance by others," Saeed develops a relationship with a black Christian preacher's daughter, showing that love can develop despite national, racial, and religious differences (219). The preacher himself counters traditional notions of love in marrying "a woman from Saeed's country"—a woman who is the mother of his daughter, Saeed's lover (219). Similarly, love as it emerges in Nadia's life counters conservative notions of what comprises a beautiful union in that she develops a sense of sexual liberation and eventually a relationship with another woman. Nadia first feels romantic affection for another woman in remembering "the girl from Mykonos" who helps her and Saeed to find and travel through a door to London (171). She dreams of returning to Mykonos, and in so doing, she feels "alive, or alarmed, regardless changed" (171). Although she never returns to realize a relationship with the woman, she does develop a relationship with "the head cook from the cooperative" in California, a "handsome woman with strong arms" who makes Nadia feel thrilled by "being seen by her, and seeing her in turn" (218). Finally, in another vignette, Hamid complements these examples of love that migration and open-mindedness allow for with an example of another similar kind of love. Much as Nadia and the cook develop a love that counters traditional perspectives that oppress women and thrives across national divides, an elderly man and a wrinkled man who emerges through a door in Amsterdam counter notions of love that function to keep them apart. They initially transcend the limits of a language barrier that divides them and eventually engage in their first kiss, challenging ideas that homosexual relationships between men or relationships between elderly individuals are somehow less valid than those between young heterosexual lovers.

Notably, Hamid's novel, too, has a capacity to achieve through the more traditional medium of language printed in a book something more than images on a screen can achieve in an image-driven digital age that is defined or perhaps even plagued by media saturation. Hamid gestures toward the limits of circumscribed, screen-based digital photographs as aesthetic objects in the digital age in his depiction of the fictionalized digital photograph that the war photographer captures of the two elderly gay men engaged in their first kiss. In the world of Hamid's fiction, the photographer creates an aesthetic, digital representation of a loving, human connection in taking the photograph—a connection that might afford new perspective on love much as Cohen's photographs afford new perspective on time, space, and the possibilities of digital art. Yet the digital photograph may live an online life as a source of controversy at best or as the object of fierce bigotry at worst.

In seeing the beauty in the connection between the two men and in what Hamid's narrator terms "a gesture of uncharacteristic sentimentality and respect," the photographer deletes the photograph, signaling the kind of transience of digital objects that I discuss in relation to "The (Former) General in His Labyrinth" (176). The photographer's action shows that beautiful connections perhaps should remain sheltered from the hate speech that an online context might foster, at least until a more socially just moment in history—a moment when viewers might sustain a perspective of the sort that Cohen's photography attempts to teach through a less politically charged subject. As a result, evidence of the connection between the two men and of the photograph as an aesthetic object that can foster perspective remains only in language. It remains off screen and evident only in the relatively safer and richer frame of Hamid's text, a potential catalyst for social change that may well have the capacity to change more minds in more meaningful ways than digital objects in the vast online world that it complements.

Hamid ends his novel by merging apparently disparate aesthetic approaches that simultaneously idealize and complicate possible resolutions to the migration crisis. In part, the conclusion of Hamid's novel resembles a stereotypical happy ending that characters in and readers of many literary novels may find elusive. It portrays a happy ending in showcasing a near-impossible reunion—or reconnection—between Saeed and Nadia in the aftermath of their exit west as an act of migration. Against all odds, they reconnect in real as opposed to virtual life at a café in the unnamed city after half a century—after "the fires" that prompted their initial exit west "had burned themselves out" (229). Their reconnection speaks to an aesthetics of connectivity that Hamid employs in the paragraphs that describe their reunion. Each of the first four paragraphs comprises a single and notably long sentence that expresses connectivity by way of its clauses, commas, and conjunctions—by way of mechanics that make connections between words possible. Yet Hamid complements this idealized and near-impossible reunion and the connectivity that he portrays through his prose style with a rhetoric of ambiguity and uncertainty. He echoes the rhetoric he employs at the end of *The Reluctant Fundamentalist*, when readers are left to wonder whether Changez, the narrator, is a terrorist or, more likely, the victim of mass xenophobia. In describing the unnamed city at this future historical moment, he notes that it "was not a heaven" but also "was not a hell" (229). He also notes that it "was familiar but also unfamiliar." Most notably, he observes that the future of Saeed and Nadia's relationship remains uncertain, as does the possibility of them ever even seeing each other again. Saeed tells Nadia that if he ever has an "evening free," he will take her to see the stars as they appear in "the deserts of Chile"—a trip they could never make in a lone evening without the help of magical doors that can only exist in the world of the novel (231,

230). But the narrator observes that when they part ways, they lack a clear sense of whether "that evening would ever come" (231). They lack a clear sense of whether they could see in real life the type of night sky that Cohen's photographs as digital art objects portray on a screen.

The ways in which Hamid merges the aesthetics of a fictionalized happy ending and the aesthetics of ambiguity and uncertainty into a hybrid whole gestures toward the notion that his novel can accomplish what lived history has yet to realize. Indeed, for Hamid, the novel as a form has more to give because it has the capacity to adapt to the changing and increasingly screen-based digital times to help humanity navigate life's questions and problems just as well as digital technology can. And Hamid shows that hybridity—the mixture or interconnection of apparently disparate things—can exist in art in fruitful ways that remain relatively elusive in the world about which art is made, despite mass migration that connects people of disparate social identities with one another and despite the connectivity that globalization and digital technology promise but often fail to fully deliver. Ultimately, fruitful hybridity in a literary work such as *Exit West* may not manifest in lived reality for Hamid's readers. Although they may become "deterritorialized" as Peter Morey imagines that readers of *The Reluctant Fundamentalist* become—readers who "expand and defamiliarize our own imaginative territory" and "find a space *between* (or at least from which to try to encapsulate) conflicting interests and positions"—they may not find themselves among like-minded individuals in their nations (136, 138, 138). A literary vision of hybrid interconnection may also not lead citizens of the globe to social responsibility toward migrant Others— toward what Bauman calls "a dialogue that aims if not at an unconditional agreement, then surely at mutual understanding" (113). But possibilities for connectivity and fruitful hybridity through interconnection as Hamid represents them in literature may lead Hamid's readers to look "for signs of hope and optimism in the future" instead of continuing to look mindlessly at the screens that surround them, to quote Hamid's words from a 2017 *New York Times* interview with Alexandra Alter ("Global Migration Meets Magic"). They might lead his readers to connect with one another through digital and nondigital means in the hope that bigotry, not people, emerges as displaced at some future time. They likewise might lead his readers to imagine and hope for what an alternate reading of his title suggests is possible: an exit of the West, or at least its destructive nativist ideology, so that a future of robust human interconnection of the sort that both Smith and Hamid celebrate can screen out and perhaps even supplant the weak digital connections that presently define and shape the segregated digital times.

CONCLUSION

Flat-World Fiction and the Textured Future

The chapters of this book showcase ways in which twenty-first-century American authors of mainstream literary fiction and authors writing about the United States negotiate with false notions that technological progress equals American or human progress in what Thomas L. Friedman terms the period of Globalization 3.0. They also negotiate with the increasing prevalence of digital devices and media, both of which metaphorically digitize humanity and which Friedman sees as metaphorically flattening the world. The works of fiction that these authors produce track changes in human relationships with emergent technologies in accord with Martin Heidegger's discussion of relationships in the opening of "The Question Concerning Technology." Specifically, Gary Shteyngart and Kristen Roupenian explore the ways in which human relationships with digital devices and media mediate and morph romantic relationships, inhibiting understanding between lovers from different cultural backgrounds and thereby love. Thomas Pynchon and Jennifer Egan consider how human relationships with digital devices and media create nostalgia for the Cold War era, alter notions of history and time, and fuel a desire for a counterdigital and countercorporate future. Joshua Ferris and Jonathan Safran Foer portray the relationship between the digital world and notions of the divine, intimating that politically charged ideas associated with religious faith and religious texts can inform contemporary approaches to understanding all texts and engaging in the digitizing world in mature, responsible, and ethical ways. Don DeLillo and Dave Eggers see the relationship between deified American capitalism and the similarly deified digital world as producing dehumanizing conditions and superficial values that mystery, agency, and engagement can counter. Finally, Zadie Smith and Mohsin Hamid examine the ways in which digitization creates the illusion of a globalized world in which individuals can easily transcend national borders even though it actually creates a world in which humans may metaphorically screen one another out.

In this conclusion, I draw attention to the ways in which flat-world fiction texturizes literature, humanity, and life in a digital age. In other words, it adds some semblance of depth to them despite or perhaps because of an *intensification* (to reference Jeffrey T. Nealon's term) of Fredric Jameson's notion of depthlessness and because of the loss of and apparent desire to regain modern depth, which is texture's implicit theoretical antecedent but which, I suggest, cannot be resuscitated in its original form after the end of modernism in a screened and depthless world. I thereby show how this literature's function stands in stark contrast to the function of the webbed, digital world that Friedman romanticizes. Notably, this fiction affords texture in the face of a decline in print culture and a purported rise of illiteracy that Shteyngart spotlights in *Super Sad True Love Story*. Works such as Sven Birkerts's *The Gutenberg Elegies* initiated conversations about this literary apocalypse in the 1990s as screen culture increasingly revealed its monolithic potential. More recently, Nicholas Carr, building on "Is Google Making Us Stupid?", his 2008 *Atlantic* article, speaks to a similar concern in *The Shallows: What the Internet Is Doing to Our Brains*. As Carr argues, "For the last five centuries, ever since Gutenberg's printing press made book reading a popular pursuit, the linear, literary mind has been at the center of art, science, and society," but this mind "may soon be yesterday's mind" due to overreliance on the internet (*The Shallows* 10). Building on notions of the demise of books and reading such as these, Farhad Manjoo argues in "Welcome to the Post-Text Future," his 2018 *New York Times* article, that words exist as futile in the contemporary world. As Manjoo puts it, "The defining narrative of our online moment concerns the decline of text, and the exploding reach and power of audio and video." As Manjoo continues, "we have only just begun to glimpse the deeper, more kinetic possibilities of an online culture in which text recedes to the background, and sounds and images become the universal language." Tacitly referencing biblical apocalypticism from the Book of Daniel, he concludes: "For text, the writing is on the wall."

In the face of claims that books, reading, and even words themselves are dead, the fictional works I consider in this study tacitly or explicitly texturize their readers' understanding of literacy and textuality, the former of which Harvey J. Graff in *The Legacies of Literacy: Continuities and Contradictions in Western Culture and Society* notably defines as akin to the kinds of digital innovations that animate these authors' literary imaginations. In Graff's words, literacy is "above all a *technology* or set of techniques for communications and for decoding and reproducing written or printed materials" (4). Hence textual fictional works that cultivate literacies function as technological and political tools in society much as digital devices and media do, as evidenced by Colin Lankshear and Peter L. McLaren's remarks in their introduction to *Critical Literacy: Politics, Praxis, and the Postmodern*. As Lankshear and McLaren frame it, "We are now much more aware than previously of the nature and role of extant literacies within established configurations of power and advantage, of centers and margins, and how literacies impact on the satisfaction of human needs and interests" (4). Lankshear and McLaren continue, observing that we "understand more fully, but still imperfectly, how literacies are implicated in the shaping of human subjects, the ideologies they bear, and their placement within social hierarchies." Thus, we more completely comprehend the ways in which literacies shape hierarchies of the kind that, for example, Shteyngart in *Super Sad True Love Story*, Roupenian in "Cat Person," DeLillo in *Cosmopolis*, Smith in "Meet the President!", and Hamid in *Exit West* address head-on.

Moreover, the authors in this study, most explicitly Ferris through his references to Wikipedia and his discussion of identity theft, put a premium on exposing information literacy as a type of literacy that digital-age Americans need in order to participate responsibly in textured contemporary conversations, particularly political ones. These writers show a hyperawareness of their roles in the production of fiction. As demonstrated by the idealizing text messages between lovers in Roupenian's "Cat Person" or by the corporate marketing strategies of the Circle in Eggers's *The Circle*, they, too, show awareness of the different ways in which fiction exists as a flexible and textured medium that is open to manipulation beyond the bounds of literary circles and the typical rhetorical situations through which mainstream literary fiction moves. They show an awareness of the notion that fiction is outright produced to masquerade as nonfiction in the digital age, a point toward which Pynchon gestures through the possibility of conspiracy in *Bleeding Edge* and a point that Dana L. Cloud explicates in *Reality Bites: Rhetoric and the Circulation of Truth Claims in U.S. Political Culture*. According to Cloud, the proliferation of fake news during the 2016 presidential election and other similar phenomena—for instance the belief that there exist "alternative facts" of the sort that U.S. counselor to the president Kellyanne Conway mentioned in her 22 January 2017 interview on *Meet the Press*—help set the terms for contemporary, media-saturated rhetorical situations ("Interview with Chuck Todd"). And as demonstrated by the 2016 election of Donald Trump, who is notably referred to as America's first Twitter President,[1] the manipulation of fiction into fact via digital means and thus the manipulation of these rhetorical situations has political consequences that are dire for people of color and hybrid Others in the United States. They have dire consequences for American society, which emerges as bifurcated in accord with DeLillo's representation of Manhattan in *Cosmopolis*. Like DeLillo's city, which is divided between largely white elites who live a life of leisure in white limousines and working-class individuals who are often immigrants and people of color who serve these elites, the United States is comprised of haves and have-nots.

Furthermore, the authors in this study recognize the need for texturized literacies that add complexity to recent conversations about approaches to reading. They present depth in what digitization is, can be, and can do, and in doing so, they showcase new modes of reading within their literary works and also open their works to these new modes of reading, for instance hyper reading as N. Katherine Hayles explores it in *How We Think* (11); surface reading as Stephen Best and Sharon Marcus theorize it in "Surface Reading: An Introduction"; and reading with as opposed to against the grain as Timothy Bewes imagines it in "Reading with the Grain: A New World in Literary Criticism." For Hayles, hyper reading "includes skimming, scanning, fragmenting, and juxtaposing texts" (*How We Think* 12). For Best and Marcus, surface reading takes "surface to mean what is evident, perceptible, apprehensible in texts; what is neither hidden nor hiding; what, in the geometrical sense, has length and breadth but no thickness, and therefore covers no depth" (9). For Bewes, reading with the grain "would not look like anything, since, in its ideal form, it would amount to thinking the text inseparably from the text," or, to employ the term to which I have consistently returned in this study, flattening criticism into the literary work (24).

By contrast, however, the authors of literary fiction upon whom I focus highlight the need for reinvigorating close reading as Jessica Pressman does in *Digital Modernism*, a book that positions close reading as a tool for readers of digital texts. More to the point, these authors aim to underscore the value of symptomatic, historicized reading in the flat world. They recognize, as I have argued, that the flat surface is highly political. In other words, to again turn to Nealon's key term for distinguishing post-postmodernism from postmodernism, this fiction invites and values the *intensification* of what Jameson's *The Political Unconscious: Narrative as a Socially Symbolic Act* theorizes as "political interpretation," the "absolute horizon of all reading and all interpretation" (Jameson 17). This fiction recognizes twenty-first-century American life's increasingly polemical political realities, which a 2014 Pew Research Center survey speaks to in finding that "Republicans and Democrats are more divided along ideological lines—and partisan antipathy is deeper and more extensive—than at any point in the last two decades" ("Political Polarization"). This fiction puts a premium on inviting readers to see value in moving between the surface and hidden depths. It exposes the tension between what is superficial and what Jameson characterizes as deep or repressed elements of the text that remain "unrealized in the surface of the text" (*The Political Unconscious* 48).

As a result, these authors either explicitly or tacitly laud multiliteracies, which, according to the New London Group in "A Pedagogy of Multiliteracies: Designing Social Futures," "broaden" conventional understandings "of literacy and literacy teaching and learning to include negotiating a multiplicity of discourses" (61). For the New London Group, there exist "two principal aspects of this multiplicity" (61). To prevalent scholarly attention, this multiplicity involves what the New London Group terms "the burgeoning variety of text forms associated with information and multimedia technologies"—the new and thereby metaphorically glossy technologies that digital humanities scholars rely on to reinvigorate the humanities (61). This multiplicity involves the text forms that all the novels and short stories in this study reference, negotiate with, challenge, and build on in accord with Hayles's observation that "print and electronic textuality deeply interpenetrate one another" ("The Future of Literature" 181). Thus, it somewhat paradoxically involves a rejuvenation of alphabetic print textuality that Anne Frances Wysocki gestures toward in "with eyes that think, and compose, and think: ON VISUAL RHETORIC," which argues that words exist as visual and aesthetic phenomena on a page or on a screen because they involve design elements. In other words, to reference the title of Cheryl E. Ball and Colin Charlton's contribution to *Naming What We Know: Threshold Concepts of Writing Studies*, this multiplicity involves the recognition that "All Writing is Multimodal," as evidenced, for instance, by the distinctive typefaces of social media messages in Shteyngart's *Super Sad True Love Story* or by the appearance of words in PowerPoint form in Egan's *A Visit from the Goon Squad*.

In addition and arguably to less attention, the multiplicity that the New London Group describes also involves the "culturally and linguistically diverse and increasingly globalized societies" that Donna Haraway anticipates in "A Cyborg Manifesto" and that are of consistent interest to the authors I include in this study (New London Group 61). It involves a hybridity that complements and texturizes technologically oriented notions of it, for example Andy Clark's notion of "cognitive hybridization" in *Natural Born Cyborgs*; Egan's conceptualization of literal hybridization through technological implantation in "Black Box"; and Smith's notion of the hybridization of physical reality through engagement with virtual reality in "Meet the President!" (Clark 4). For the authors I address, hybridity is not just a digital-age concept. It is the result of migration and its aftereffects, a point to which Shteyngart and Hamid explicitly draw attention in their respective novels and a point that Homi K. Bhabha famously theorizes in *The Location of Cul-*

ture, an inadvertent extension of Haraway's theoretical work. For Bhabha, to be hybrid is to be "neither the one thing nor the other" (49). Hence hybridity involves the embodiment of liminal identity that allows for multiliteracy as a way of reading and understanding the diverse world. Likewise, intersectionality as Patricia Hill Collins and Sirma Bilge theorize it speaks to Bhabha's notion of hybridity.[2] For Collins and Bilge, intersectionality involves the intermingling of identities. It entails the intersection of race, class, gender, sexuality, nationality, ethnicity, religion, and linguistic heritage as well as less considered identities including, for instance, professional identity, which Ferris shows particular interest in in *To Rise Again at a Decent Hour*, or familial identity, which Pynchon attends to in *Bleeding Edge*. It likewise involves the identities we develop as a result of formative experiences and traumas, for instance the survivor identity that the kleptomaniacal Sasha Grady develops as a result of witnessing her father abuse her mother in her youth in Egan's *A Visit from the Goon Squad*. Significantly, too, intersectionality also comes to involve the online identities that users of digital media develop in the digital age: remediated, screen-based identities that emerge as related to other noteworthy identities and intersectional in the same way.

The authors examined in this book suggest that hybrid and intersectional identities and the hybrid culture they produce add texture to contemporary conversations involving social justice, in ways a quite simple term that historically involves activist work in the streets for the underprivileged and the subversion of structural inequality, but a term that paradoxically comes to lack clear meaning in the twenty-first century, perhaps because of its recent widespread use or perhaps because everyday slacktivist acts have come to stand in for increasingly rare and authentically transformative activist acts. As a result of the persistent post-9/11 anxieties and rhetorics to which Pynchon and Egan in particular allude and also in the face of Trump-era hostility toward international and hybrid Others under the guise of making America "great again," there exists an apparent push to revitalize American exceptionalism, as the title of Henry Luce's *Life* magazine editorial famously speaks to it. As evidenced by the increasing number of flat earthers, believers in the notion that the world is literally flat who cite evidence presented by a YouTube video as the foundation for their beliefs, ignorance has gone viral.[3] This ignorance has a relationship with the illiteracy that white or other privileged Americans may demonstrate toward the identities of Others, which takes the form of xenophobia, racism, sexism, ethnocentrism, and violence. In a 2018 *Guardian* article, Hamid describes an unsettling "global trend" to attack the de facto cultural hybridity that defines the United States and the world in our so-called interconnected, globalized times ("Mohsin Hamid on the Rise of Nationalism"). As Hamid articulates it, "All around the world, governments and would-be governments appear overwhelmed by complexity and are blindly unleashing the power of fission, championing quests for the pure." He elaborates, observing that in "India a politics of Hindu purity is wrenching open deep and bloody fissures in a diverse society. In Myanmar a politics of Buddhist purity is massacring and expelling the Rohingya." The problem also exists in the United States, where, according to Hamid, "a politics of white purity is marching in white hoods and red baseball caps, demonising Muslims and Hispanic people, killing and brutalising black people, jeering at intellectuals, and spitting in the face of climate science," thereby resulting in ecological devastation as Smith portrays it in "Meet the President!" ("Mohsin Hamid on the Rise of Nationalism").

In turn, the authors considered in this book exemplify an impulse to support hybrid and intersectional identities and hybrid culture through fictional works that pointedly texturize and politicize digitization and view it as part of human identity, a reality that Ferris spotlights by naming the smartphonelike devices of *To Rise Again at a Decent Hour* "me-machines" (74). Like Brian T. Edwards and Dilip Parameshwar Gaonkar, who in their introduction to *Globalizing American Studies* propose that the field of American Studies should free itself of "the long and enduring myth of American exceptionalism" that Luce's article propagates, these authors see freedom from that imperialist perspective as essential to the realization of real social concern for hybridized humanity (5). They see it as necessary for the realization of a social justice that, for instance, undermines the super-rich or empowers people of color and women in sleek digital times through uncomfortable processes in and beyond the streets. Their fictional works run counter to Friedman's argument, in *The World Is Flat*, that digitization can heal the injustices of the nation or the globalized world.

Indeed, through their fictional works, these authors question utopian characterizations of digitization such as that of Friedman, who observes that "it is now possible for more people than ever to collaborate and compete in real time with more other people on more different kinds of work from more different corners of the planet on a more equal footing than at any previous time in the history of the world—using computers, e-mail, networks, teleconferencing, and dynamic new software" (*The World Is Flat* 8). They implicitly question the politics of accelerationists, who, as Andy Beckett explains in a 2017 *Guardian* article, "argue that technology, particularly computer technology, and capitalism, particularly the most aggressive, global variety, should be massively sped up and intensified—either because this is the best way forward for humanity, or because there is no alternative." They question whether digital connectivity inherently leads to love, countercorporate values, transcendence, knowledge, empathy, or individual or collective empowerment. In short, they suggest that the ever-emergent digital world has outrun the production of philosophy that might help humanity use digital tools to live more genuinely connected, empathic, and socially concerned lives. It has outrun the development of philosophy that might imagine the world according to socially progressive as opposed to technocratic or corporate-born definitions of American progress through means that, by definition, must involve messy disruption of a problematic social order.

Instead of presenting romantic notions of digitization, these authors present poignant portraits of often hybrid fictionalized Americans and an often tragic America and world, texturizing and accentuating the thorny beauty of reality that fiction can capture. These portraits stand out as particularly moving because their authors juxtapose them with the digital world, which, at its worst, is sterile and lifeless in accord with Benny Salazar's view of digital music in Egan's *A Visit from the Goon Squad*. For instance, Gary Shteyngart gives texture to the beauty of meaningful love between hybrid American Others in *Super Sad True Love Story*. Roupenian highlights the complicated nature of women's sexual experiences in patriarchal America in "Cat Person." Pynchon shows the beauty of feminist sisterhood and motherhood in *Bleeding Edge*. Egan spotlights the raw energy of old music played live on instruments in basements or on textured records in *A Visit from the Goon Squad*. Ferris portrays the enriching experience of engaging in humanitarian work at home and overseas in *To Rise Again at a Decent Hour*. Foer dramatizes the knotty realities of maturation, alienation in old age, and death in *Here I Am*. DeLillo showcases the mystique of old New York City neighborhoods such as the Diamond District and Hell's Kitchen in *Cosmopolis*. Eggers renders the deep, natural world of the San Francisco Bay in *The Circle* as having transcendent qualities. Smith captures the musty feel of communities of diverse locals who come together to find spiritual meaning in life in the face of life's end at the conclusion of "Meet the President!" And Hamid portrays the instability inherent in modern acts of migration and the feeling of community within the textured communal living situations that migrants develop in a globalized world in *Exit West*.

Ultimately, Shteyngart and Roupenian, Pynchon and Egan, Ferris and Foer, DeLillo and Eggers, and Smith and Hamid write fiction that helps readers navigate the increasingly hybrid times to imagine new and philosophically informed approaches to using digital devices and media in and beyond American borders. These authors invite readers to push beyond circumscribed understandings of texts and contexts. They texturize digital-age notions of identity, pushing readers beyond the limits of their filter bubbles, to reference a concept that Eli Pariser introduces in *The Filter Bubble: How the Personalized Web Is Changing What We Read and How We Think*.[4] They give texture to the ways in which readers imagine hybrid Others to themselves in the face of ubiquitous literal and metaphorical screens that screen individuals from one another and that screen America from a future beyond the haughty ideals of the American Century. In the process, these authors capture a literary snapshot of the peculiar historical moment in which Americans and global citizens live. In this moment, they feel enthralled by the magic of digitization, uneasy about notions of progress toward the future, increasingly and unsettlingly disconnected from one another, and perhaps ineffectual at igniting social change in a physical world that emerges as a new kind of alien in flat-world fiction as a new type of science fiction.

As Sherry Turkle suggests in *Reclaiming Conversation*, digital natives no longer feel comfortable engaging with one another in physical circumstances. They no longer feel at ease talking to one another with spoken words or in person, a reality that the Covid-19 crisis exacerbates at the end of the period in American literary history that this book examines. But flat-world fiction as an aesthetic medium composed through the building blocks of imagination can begin an important conversation. It can help readers imagine new futures in which they texturize their own utility in the world. And it can counter the superficial beauty of flat, sleek things in digital times with the notion that, to borrow Hamid's words from "The Great Divide," a *New York Times Magazine* essay about the online and real walls that divide us, there "is magic" in a hybrid, "mongrelized society." There is magic in contemplating our own hybridities and potential heterodoxies and in transcending the bounds of identity and nations through hybrid means that embrace the interplay of the digital and the real. And there is magic in generating a counternarrative to corporate-born globalization in the future through the meaningful understanding that nuanced, textured, and textual fiction enduringly delivers.

NOTES

Introduction. American Literature and Digital Technology in the New Millennium

1. See "The American Century," Luce's 1941 *Life* magazine editorial.
2. In the twentieth century, Leslie Fiedler, John Barth, Alvin Kernan, Sven Birkerts, and others prophesied the demise of print books, and twenty-first-century authors have made similar claims. As Jonathan Franzen puts it in "The Reader in Exile," "For every reader who dies today, a viewer is born" (165).
3. Consider, for instance, Cass R. Sunstein's *Infotopia: How Many Minds Produce Knowledge*, which celebrates the cumulative knowledge that develops as a result of the internet, or Clay Shirky's *Cognitive Surplus*, which lauds new media for their ability to connect disparate individuals and allow them to collaborate.
4. Landrum et al.'s paper, "YouTube as a Primary Propagator of Flat Earth Philosophy," was delivered at the 2019 Annual Meeting of the American Association for the Advancement of Science in Washington, DC.
5. Among the most noteworthy digital developments of the Cold War era was the first modern computer, arguably ENIAC, which Presper Eckert and John Mauchly completed in November 1945 and operated at the University of Pennsylvania.
6. As Jen Schradie explains in *The Revolution That Wasn't: How Digital Activism Favors Conservatives*, activism seems "cheap, accessible, fast, and open to all" in its online incarnation, but in reality, it fails to offer "a quick technological fix to repair our broken democracy" and reproduces or even intensifies "preexisting power imbalances" (ix, 7, 7).
7. Notably, scholars such as Val Dusek have addressed debates about the newness of different technologies. According to Dusek in *Philosophy of Technology: An Introduction*, the "technology as hardware" definition suggests that digital tools are like tools that humans have used throughout history to define their humanity (31). By contrast, scholars of technology might see ways in

which new technology constructs distinctively new life experiences. Thus, they view technology as "rules rather than tools," focusing on human behavior rather than on the inventions themselves (Dusek 32). Alternately still, they may view technology as a system that involves "people and organizations, productive skills, living things, and machines" (Dusek 35).

8. See *The Education of Henry Adams*.

9. In *Post-Postmodernism, or, the Logic of Just-in-Time Capitalism*, Jeffrey T. Nealon sees the current post-postmodern era as defined by an "intensification and mutation within postmodernism" (ix).

10. As Brian Attebery explains, the "magazine era" of science fiction, which lasted from 1926 to 1960, shaped much science fiction that followed it, and the location of most science fiction publishers prompted a strong "association between" science fiction and "American culture" (32).

Chapter 1. Relationships with Technology in Gary Shteyngart's *Super Sad True Love Story* and Kristen Roupenian's "Cat Person"

1. Shteyngart was born in Leningrad in the Soviet Union (now St. Petersburg, Russia) in 1972 and emigrated to the United States at age seven with his parents, who settled in New York City. His first language is Russian, but he writes in English. He is married to Esther Won, a Korean American lawyer.

2. As Shteyngart explains in an interview with Elizabeth Tannen, he wants, via the fiction he writes, "to explain the world to [him]self," and the world as he sees it is shaped by digital and social media ("A Conversation with Gary Shteyngart" 178).

3. For Hutcheon, parody is not a "ridiculing imitation of the standard theories and definitions that are rooted in eighteenth-century theories of wit," but "that seemingly introverted formalism" which functions to "enshrine the past" and "question it" (26, 22, 126, 126).

4. For instance, in *Little Failure*, Shteyngart describes his own early life experiences of admiring the "informal and direct" ways in which Americans around him spoke English "with the words circling the air like homing pigeons" (96). He includes Russian transliterations and examples of Russian-immigrant English in *Little Failure*.

5. For instance, N. Katherine Hayles argues that the digital age has transformed reading because it has brought about three kinds of reading: "close, hyper-, and machine" (*How We Think* 11). Harriet Griffey's 2018 *Guardian* article, "The Lost Art of Concentration," suggests that digitization inhibits human concentration and thus the ability to read.

Chapter 2. Searching History in Thomas Pynchon's *Bleeding Edge* and Jennifer Egan's *A Visit from the Goon Squad*

1. For Wajcman, "our visions of the future" involve "acceleration" (9).

2. See David Cowart's "Thirteen Ways of Looking: Jennifer Egan's *A Visit from the Goon Squad*," which suggests that "Egan's perspectivism [...] transforms that of the Cubists or Wallace Stevens" (245).

3. My reading counters that of Joseph Darlington in "Capitalist Mysticism and the Historicizing of 9/11 in Thomas Pynchon's *Bleeding Edge*," who argues that "the principled yet violent

and unflinching qualities of Nicholas Windust are the very principles that Pynchon attributes to the mythical figure of the technophobic King Ludd in his 1984 essay 'Is It Okay to Be a Luddite?'" (251). Darlington thereby argues that Windust is the badass of Pynchon's novel.

4. Chapter 4, "Safari," is set in the early 1970s and depicts the earliest events of the book, and chapter 13, "Pure Language," is set in the early 2020s, the near future, and depicts the latest events of the book.

Chapter 3. The Digital Divine in Joshua Ferris's *To Rise Again at a Decent Hour* and Jonathan Safran Foer's *Here I Am*

1. As Ferris explains in "A Conversation with Joshua Ferris" in the Reading Group Guide for *Then We Came to the End*, "initial versions of my novel were top-heavy with complaint," but "after a certain remove," he realized "that office life has a lot to offer: companionship, the opportunity to clown around, lunch mates" (6–7).

2. Foer interweaves fictionalized nonfictional narrative by Alex Perchov, magical-realist narrative of Holocaust-era history by Jonathan Safran Foer (the character), and letters written to Foer (the character) by Perchov. Also, Foer's mother, Esther Safran Foer, is the child of Holocaust survivors who immigrated to the United States.

3. As Winthrop puts it in "A Model of Christian Charity," referencing Jesus's words from Matthew 5:14, the settlers of the New World would "be as a city upon a hill" (216).

4. According to McGrath and McGrath, Richard Dawkins makes his argument with "total dogmatic conviction" that aligns him "with a religious fundamentalism that refuses to allow its ideas to be examined or challenged" (14).

5. According to Karl Barth in *The Epistle to the Romans*, God is wholly Other and not "known or knowable" (36).

6. See *The Post-Truth Era: Dishonesty and Deception in Contemporary Life*, in which Keyes suggests that the "World Wide Web is a mishmash of rumor passing as fact, press releases posted as news articles, deceptive advertising, malicious rumors, and outright scams" (205).

7. Cotton Mather's 1693 book, *The Wonders of the Invisible World*, suggests that Satan operates in the world and influences metaphysical occurrences in Salem, Massachusetts.

8. For Weber's discussion of the interplay between Protestantism and capitalism, see *The Protestant Ethic and the "Spirit" of Capitalism and Other Writings*. For a consideration of "market fundamentalism" as a dominant doctrine in the United States, see Malise Ruthven's *Fundamentalism: A Very Short Introduction* (21).

9. According to S. Todd Atchison, "The piecing together" of "fragmentary" narratives in *Extremely Loud and Incredibly Close* "forces readers to actively engage with the text. Such activity reveals not only how readers approach the text but also how we experience the other" (360, 360, 361–362).

10. As Gleich suggests, *Falling Man*, *Man in the Dark*, and *Extremely Loud and Incredibly Close* provide "readers with a space for beginning to think about an ethics that can respond to the aesthetics of an age dominated by images" (163). According to Vanderwees, *Extremely Loud and Incredibly Close* also "prompts readers to consider ethical relationships between themselves and others" (177).

11. Foer wrote *All Talk* for HBO but opted against airing any episodes of it.

12. Foer won the National Jewish Book Award for *Everything Is Illuminated* in 2001, and Jacob wins it in fiction in 1998.

13. In a 2016 NPR interview, Foer tells Gross: "Ritual has become more important to me as I've gotten older. It's not always religious ritual but it often borrows from Judaism." And as he continues to Gross, "Most of the times that I think about my relationship to Judaism, I not only accuse myself of a shallowness, but I feel certain that there's a shallowness there. That's not a bad thing, really. You know, to acknowledge a shallowness implies a kind of aspiration—like, a hope to have more depth."

14. According to Codde, in his first and second novels, Foer "suggests that pictures can reveal more about the past than the rich variety of verbal records that feature in the novel, though he seems to become increasingly pessimistic about their testimonial value" (249).

Chapter 4. Cybercapitalism in Don DeLillo's *Cosmopolis* and Dave Eggers's *The Circle*

1. For instance, Tom Junod notes that DeLillo "has been writing the post-9/11 novel for the better part of four decades, and his *pre*-9/11 novel, the magnum-opusy *Underworld*, was prescient enough to put the looming towers on its cover, standing high and ready to fall" ("The Man Who Invented 9/11").

2. I build on Tom LeClair's classic argument that humanity is part of "ecological systems" that include corporate and technological forces in DeLillo's work (xi). Also, I build on Randy Laist's more recent claim that "DeLillo is a phenomenologist of the contemporary technoscape and an ecologist of our new kind of natural habitat" (3).

3. Co-founded by Eggers and Lola Vollen, Voice of Witness is a nonprofit organization that publishes nonfiction stories about human rights abuses. Co-founded by Eggers and Nínive Clements Calegari in 2002, 826 Valencia is a nonprofit organization that provides tutoring to students.

4. As Laist observes, "DeLillo's characters are cyborgs manqué, devotees of a technological transcendence that is tantalizingly immanent in the glistening precision of consumer technologies" (4). And as Aristi Trendel notes in a discussion of *Cosmopolis*, "Eric Packer is a hybrid, a human morphing into a posthuman" (117).

5. According to the narrator, Brutha Fez is hybrid in that he mixes "languages, tempos and themes" (*Cosmopolis* 134). For a discussion of Brutha Fez's postcolonial hybridity, see Rolf J. Goebel's "From Postcoloniality to Global Media Culture."

6. As Benno explains while wondering aloud whether he is forty-one or forty-two, "I don't keep track, because why should I?" (*Cosmopolis* 189).

7. According to the twin paradox, if one twin travels through space near or at the speed of light on a fast-paced rocket, that twin ages more slowly than the twin on earth.

8. Eric unravels because of the yen; Benno unravels because of the baht.

9. For instance, see Nicholas Bakalar's 2018 *New York Times* article, "How Nighttime Tablet and Phone Use Disturbs Sleep."

10. For instance, Sven Birkerts prophesied the death of books in *The Gutenberg Elegies*, and

Jonathan Franzen expressed anxiety about the demise of print books and reading in the face of digitization in "The Reader in Exile."

11. See John A. McClure's "DeLillo and Mystery," which argues that "DeLillo's work urges the reader to perform a discrimination of mysteries—to check his or her fascination with forensic and esoteric mysteries and explore the possibility of apophatic and sacramental modes of being" (167).

12. The Manx Martin sections progress in a linear fashion, but the story of Nick Shay runs in reverse.

13. As Kinski asks Eric, "Why die when you can live on a disk? A disk, not a tomb. An idea beyond the body" (*Cosmopolis* 105).

14. Describing Eric's desire to return to his dead father's old neighborhood, the narrator observes, "He wanted to feel it, every rueful nuance of longing" (*Cosmopolis* 159).

15. As Kinski puts it, capitalists believe in "[e]nforced destruction. Old industries have to be harshly eliminated. New markets have to be forcibly claimed. Old markets have to be re-exploited. Destroy the past, make the future" (92–93).

16. According to Hungerford, DeLillo "transfers a version of mysticism from the Catholic context into the literary one [. . .] through the model of the Latin mass" (343). As Hungerford explains, "opponents" and "advocates" of the Latin Mass have described it "in similar terms: both spoke of 'screens' and 'barriers' and lack of transparent meaning" (357).

Chapter 5. National Divides and Digitization in Zadie Smith's "Meet the President!" and Mohsin Hamid's *Exit West*

1. In *On Beauty*, Smith reimagines features of Forster's work in digital terms. For instance, she morphs the letters with which Forster's work opens into emails that Jerome Belsey sends to Howard Belsey, his father. And she subtly explores the ways in which digital technology allows for and inhibits connection, juxtaposing Howard's younger son Levi, who sustains a monkish devotion to digital music, with Howard, a traditional humanities scholar who shies away from digital culture, at least until he delivers his first PowerPoint lecture while seeing what Magdalena Mączyńska calls possibilities for "redemptive reconciliation" with his estranged wife, Kiki (Mączyńska 135). *NW*, published immediately after *On Beauty*, likewise addresses the problems and possibilities of digital-age technology. In it, characters "complain about the evils of technology" while they have "their phones laid next to their dinner plates" (*NW* 97). They live with memories of "when you turned the [phone] dial with your finger" while children speak "to videos of each other" (157). Most notably, Smith employs protagonist Natalie Blake to explore the relationship between love, sex, fantasy, reality, and the internet. Natalie visits a website where she posts and responds to personal ads that solicit risqué sex among strangers.

2. When asked in a 2005 *Atlantic Monthly* interview with Jessica Murphy Moo whether the time she spent in Cambridge influenced her choice of *On Beauty*'s setting, Smith notes that "Massachusetts absolutely inspired it" ("Zadie, Take Three").

3. According to Bergholtz, Smith in *White Teeth* "shows fundamentalism to be a woefully inadequate response to globalization" (541).

4. As Lucienne Loh suggests, "Smith's time in America provided her an opportunity to con-

template contemporary English life anew," but it also allowed her to contemplate the United States and its politics in the world (171). Indeed, in addition to setting much of *On Beauty* in the United States, Smith has written about the United States or figures from American popular culture or history in works such as "Escape from New York," a short story about Michael Jackson, Elizabeth Taylor, and Marlon Brando escaping New York City in the immediate aftermath of the 9/11 terrorist attacks, and "Crazy They Call Me," a short story about the life and death of African American jazz singer Billie Holiday.

5. In a conversation with Ruby Cutolo, Hamid says he sees himself as geographically "nomadic" (Cutolo 22). Also, for instance, Claire Chambers observes that like several other contemporary Pakistani writers, Hamid writes from a "liminal" position "between West and East" (124). Similarly, Peter Morey calls Hamid's *The Reluctant Fundamentalist* an example of "a sort of deterritorialization of literature which forces readers to think about what lies behind the totalizing categories of East and West, 'Them and Us' and so on" (138). Along the same lines, Ahmed Gamal calls Hamid a "deterritorialized" writer of "post-migratory literature" (597).

6. Bauman begins his books by gesturing toward the connection that he later explores more fully. He states that "TV news, newspaper headlines, political speeches and Internet tweets" are "currently overflowing with references to the 'migration crisis'" (1).

7. The digital divide is the notion that privileged individuals have access to digital technology that less privileged individuals lack.

8. According to Brad Buchanan, "Smith has a great deal in common with other notable theorists and critics of the 'posthuman'" (13).

9. As Winthrop expressed it on the *Arbella* in 1630, the Puritan colonies in the New World would be "as a city upon a hill," with the world watching them as an example as they prepared for the Second Coming of Christ (216).

10. In "Generation Why?" Smith discusses the concept of connection as Mark Zuckerberg refers to it, noting that Zuckerberg "uses the word 'connect' as believers use the word 'Jesus,' as if it were sacred in and of itself." As she continues, "Connection is the goal. The quality of that connection, the quality of the information that passes through it, the quality of the relationship that connection permits—none of this is important."

11. As Fukuyama explains in "The End of History?", an essay that developed into *The End of History and the Last Man*, there exists after the Cold War a "feeling that something very fundamental has happened in world history" and a sense that Western neoliberalism emerges as the "final form of human government" (3, 4). Notably, after 9/11, Fukuyama revised his argument in *Our Posthuman Future: Consequences of the Biotechnology Revolution*, which notes that "there can be no end to history without an end of modern natural science and technology" (15).

12. In *NW*, Smith tells the stories of four interconnected characters who come from Caldwell in Northwest London, yet these characters exist as disconnected from one another emotionally.

13. As Smith explains in "Elegy for a Country's Seasons," "The terrible truth is that we had a profound, historical attraction to apocalypse." As she puts it in *NW*, "There must be something attractive about the idea of apocalypse" (99).

14. Hamid's interest in the subject of globalization is evident in his earlier works, most notably in *The Reluctant Fundamentalist* and in *How to Get Filthy Rich in Rising Asia*, the latter of which Angelia Poon sees as representing "capitalist realism" (147).

15. In "Angel Gabriel," a review of Gabriel García Márquez's *Chronicle of a Death Foretold*, Salman Rushdie speaks to the political critiques that magical realism such as García Márquez's offers, writing that it "expresses a genuinely 'Third World' consciousness."

16. As Bauman puts it, "the Internet is not *the cause* of the rising numbers of morally blind and deaf internauts—but it greatly facilitates and beefs up that rise" (108).

Conclusion. Flat-World Fiction and the Textured Future

1. According to Navneet Alang in "Trump Is America's First Twitter President. Be Afraid," a 2016 article in the *New Republic*, "Trump will be the Twitter President in the worst possible way, giving a whole new meaning to the term 'bully pulpit.' And that's a shame, because his use of Twitter has been politically revolutionary in a way that could have been harnessed for the greater good."

2. As Patricia Hill Collins and Sirma Bilge suggest, "Intersectionality is a way of understanding and analyzing the complexity in the world, in people, and in human experiences" (2).

3. See Ian Sample's "Study Blames YouTube for Rise in Number of Flat Earthers."

4. Pariser's *The Filter Bubble* explains and critiques the personalized, circumscribed online experiences that internet users have due to filter bubbles that sites such as Google create. According to Pariser, "Most of us assume that when we Google a term, we all see the same results—the ones that the company's famous Page Rank algorithm suggests are the most authoritative based on other pages' links. But since December 2009, this is no longer true" (2). As Pariser puts it, "Now you get the result that Google's algorithm suggests is best for you in particular—and someone else may see something entirely different. In other words, there is no standard Google anymore" (2).

WORKS CITED

Adams, Henry. *The Education of Henry Adams*, edited by Ira B. Nadel, Oxford University Press, 1999.
Alang, Navneet. "Trump Is America's First Twitter President. Be Afraid." *New Republic*, 15 Nov. 2016, https://newrepublic.com/article/138753/trump-americas-first-twitter-president-afraid. Accessed 24 June 2019.
Allington, Daniel, Sarah Brouillette, and David Golumbia. "Neoliberal Tools (and Archives): A Political History of Digital Humanities." *Los Angeles Review of Books*, 1 May 2016, https://lareviewofbooks.org/article/neoliberal-tools-archives-political-history-digital-humanities. Accessed 18 June 2019.
Alter, Alexandra. "Global Migration Meets Magic in Mohsin Hamid's Timely Novel." *New York Times*, 7 Mar. 2017, https://www.nytimes.com/2017/03/07/arts/exit-west-mohsin-hamid-refugee-.html. Accessed 1 Sept. 2017.
Anderson, Craig A., Douglas A. Gentile, and Katherine E. Buckley. *Violent Video Game Effects on Children and Adolescents: Theory, Research, and Public Policy*. Oxford University Press, 2007.
Anderson, M. T. *Feed*. Candlewick Press, 2002.
Ariel, David S. *What Do Jews Believe? The Spiritual Foundations of Judaism*. Schocken Books, 1995.
Armstrong, Karen. *The Battle for God: A History of Fundamentalism*. Random House, 2000.
Asimov, Isaac. *I, Robot*. 1950. Bantam Books, 1991.
Atchison, S. Todd. "'Why I am writing from where you are not': Absence and Presence in Jonathan Safran Foer's *Extremely Loud and Incredibly Close*." *Journal of Postcolonial Writing*, vol. 46, nos. 3–4, 2010, pp. 359–368.
Attebery, Brian. "The Magazine Era: 1926–1960." *The Cambridge Companion to Science Fic-*

tion, edited by Edward James and Farah Mendlesohn, Cambridge University Press, 2003, pp. 32–47.

Auster, Paul. "Random Notes—September 11, 2001, 4:00 P.M.; Underground." *110 Stories: New York Writes after September 11*, edited by Ulrich Baer, New York University Press, 2002, pp. 34–36.

Bakalar, Nicholas. "How Nighttime Tablet and Phone Use Disturbs Sleep." *New York Times*, 23 May 2018, https://www.nytimes.com/2018/05/23/well/mind/how-nighttime-tablet-and-phone-use-disturbs-sleep.html. Accessed 19 Dec. 2018.

Ball, Cheryl E., and Colin Charlton. "All Writing Is Multimodal." *Naming What We Know: Threshold Concepts of Writing Studies*, edited by Linda Adler-Kassner and Elizabeth Wardle, Utah State University Press, 2016, pp. 42–43.

Barth, John. *Coming Soon!!!* Houghton Mifflin, 2001.

Barth, Karl. *The Epistle to the Romans*. 1933. Translated by Edwyn C. Hoskyns, 6th ed., Oxford University Press, 1968.

Baudrillard, Jean. *Simulacra and Simulation*. 1981. Translated by Sheila Faria Glaser, University of Michigan Press, 1994.

Bauman, Zygmunt. *Strangers at Our Door*. Polity, 2016.

Beaumont, Matt. *e: A Novel*. Plume, 2000.

———. *The e Before Christmas*. Harper Collins, 2000.

———. *e Squared: A Novel*. Plume, 2010.

Beckert, Sven, and Christine Desan. Introduction. *American Capitalism: New Histories*, edited by Beckert and Desan, Columbia University Press, 2018, pp. 1–32.

Beckett, Andy. "Accelerationism: How a Fringe Philosophy Predicted the Future We Live In." *The Guardian*, 11 May 2017, https://www.theguardian.com/world/2017/may/11/accelerationism-how-a-fringe-philosophy-predicted-the-future-we-live-in. Accessed 10 Dec. 2019.

Bergholtz, Benjamin. "'Certainty in Its Purest Form': Globalization, Fundamentalism, and Narrative in Zadie Smith's *White Teeth*." *Contemporary Literature*, vol. 57, no. 4, 2016, pp. 541–568.

Best, Stephen, and Sharon Marcus. "Surface Reading: An Introduction." *Representations*, vol. 108, no. 1, 2009, pp. 1–21.

Betts, Alexander. *Forced Migration and Global Politics*. Wiley-Blackwell, 2009.

Bewes, Timothy. "Reading with the Grain: A New World in Literary Criticism." *differences: A Journal of Feminist Cultural Studies*, vol. 21, no. 3, 2010, pp. 1–33.

Bhabha, Homi K. *The Location of Culture*. 1994. Routledge, 2004.

The Bible: Authorized King James Version. Edited by Robert P. Carroll and Stephen Prickett, Oxford University Press, 1997.

Birkerts, Sven. *The Gutenberg Elegies: The Fate of Reading in an Electronic Age*. Faber and Faber, 1994.

Blum, Andrew. *Tubes: A Journey to the Center of the Internet*. Ecco, 2012.

Boxall, Peter. *Twenty-First-Century Fiction: A Critical Introduction*. Cambridge University Press, 2013.

Boyagoda, Randy. "Digital Conversion Experiences in Don DeLillo's *Cosmopolis*." *Studies in American Culture*, vol. 30, no. 1, 2007, pp. 11–26.

Bremmer, Ian. *Us vs. Them: The Failure of Globalism*. Portfolio/Penguin, 2018.
Breslin, James E. B. *Mark Rothko: A Biography*. University of Chicago Press, 1993.
Bridle, James. *New Dark Age: Technology and the End of the Future*. Verso, 2018.
Buchanan, Brad. "'The Gift That Keeps on Giving': Zadie Smith's *White Teeth* and the Posthuman." *Reading Zadie Smith: The First Decade and Beyond*, edited by Philip Tew, Bloomsbury, 2014, pp. 13–25.
Burke, Peter. *Cultural Hybridity*. Polity, 2009.
Butler, Octavia. "Bloodchild." 1984. *Bloodchild and Other Stories*, 2nd ed., Seven Stories Press, 2005, pp. 1–32.
Caidin, Martin. *Cyborg*. Arbor House, 1972.
Carr, Nicholas. "Is Google Making Us Stupid? What the Internet Is Doing to Our Brains." *The Atlantic*, July/Aug. 2008, https://www.theatlantic.com/magazine/archive/2008/07/is-google-making-us-stupid/306868. Accessed 24 June 2019.
———. *The Shallows: What the Internet Is Doing to Our Brains*. W. W. Norton, 2011.
Chagall, Marc. *Solitude*. 1933, Tel Aviv Museum of Art, Tel Aviv.
Chambers, Claire. "A Comparative Approach to Pakistani Fiction in English." *Journal of Postcolonial Writing*, vol. 47, no. 2, 2011, pp. 122–134.
Chekhov, Anton. *Three Years*. *The Complete Short Novels*. Translated by Richard Pevear and Larissa Volokhonsky, Vintage Classics, 2004, pp. 329–432.
Clark, Andy. *Natural-Born Cyborgs: Minds, Technologies, and the Future of Human Intelligence*. Oxford University Press, 2003.
Cloud, Dana L. *Reality Bites: Rhetoric and the Circulation of Truth Claims in U.S. Political Culture*. Ohio State University Press, 2018.
Codde, Philippe. "Philomela Revisited: Traumatic Iconicity in Jonathan Safran Foer's *Extremely Loud and Incredibly Close*." *Studies in American Fiction*, vol. 35, no. 2, 2007, pp. 241–254.
Cohen, Joshua. *Book of Numbers*. Random House, 2015.
———. *Four New Messages*. Graywolf Press, 2012.
Collado-Rodríguez, Francisco. "Intratextuality, Trauma, and the Posthuman in Thomas Pynchon's *Bleeding Edge*." *Critique: Studies in Contemporary Fiction*, vol. 57, no. 3, 2016, pp. 229–241.
Collins, Patricia Hill, and Sirma Bilge. *Intersectionality*. Polity Press, 2016.
Conway, Kellyanne. Interview with Chuck Todd. *Meet the Press*. NBC News, 22 Jan. 2017, https://www.nbcnews.com/meet-the-press/meet-press-01-22-17-n710491. Accessed 24 June 2019.
Coogan, Michael. *The Old Testament: A Very Short Introduction*. Oxford University Press, 2008.
Cowart, David. "Anxieties of Obsolescence: DeLillo's *Cosmopolis*." *The Holodeck in the Garden: Science and Technology in Contemporary American Fiction*, edited by Peter Freese and Charles B. Harris, Dalkey Archive Press, 2004, pp. 179–191.
———. "'Down on the Barroom Floor of History': Pynchon's *Bleeding Edge*." *Postmodern Culture*, vol. 24, no. 1, 2013, n.p.
———. "Thirteen Ways of Looking: Jennifer Egan's *A Visit from the Goon Squad*." *Critique: Studies in Contemporary Fiction*, vol. 56, no. 3, 2015, pp. 241–254.

———. *Thomas Pynchon and the Dark Passages of History*. University of Georgia Press, 2011.
Cutolo, Ruby. "Mohsin Hamid: The Meeting of East and West." *Publishers Weekly*, 26 Nov. 2012, pp. 21–22.
Da, Nan Z. "The Digital Humanities Debacle: Computational Methods Repeatedly Come Up Short." *Chronicle of Higher Education*, 27 Mar. 2019, https://www.chronicle.com/article/The-Digital-Humanities-Debacle/245986. Accessed 18 June 2019.
Dalsgaard, Inger H. "Science and Technology." *The Cambridge Companion to Thomas Pynchon*, edited by Inger H. Dalsgaard, Luc Herman, and Brian McHale, Cambridge University Press, 2012, pp. 156–167.
Danielewski, Mark Z. *House of Leaves*. 2nd ed., Pantheon Books, 2000.
Darlington, Joseph. "Capitalist Mysticism and the Historicizing of 9/11 in Thomas Pynchon's *Bleeding Edge*." *Critique: Studies in Contemporary Fiction*, vol. 57, no. 3, 2016, pp. 242–253.
Degler, Carl. *Out of Our Past: The Forces That Shaped Modern America*. 3rd ed., Harper Perennial, 1984.
DeLillo, Don. *Cosmopolis*. Scribner, 2003.
———. "In the Ruins of the Future." *Harper's*, Dec. 2001, pp. 33–40.
———. *Point Omega*. Scribner, 2010.
———. "The Power of History." *New York Times Book Review*, 7 Sept. 1997, https://archive.nytimes.com/www.nytimes.com/library/books/090797article3.html. Accessed 19 Dec. 2018.
———. "Q&A; with Don DeLillo: The Novelist on Baseball, Technology, and How French Philosophy Has Infected the White House." Interview with Kevin Gray. *Details*, Apr. 2006, https://web.archive.org/web/20100605102309/http://www.details.com/celebrities-entertainment/men-of-the-moment/200607/wiseguy-novelist-don-delillo. Accessed 19 Dec. 2018.
———. *The Silence*. Scribner, 2020.
———. *Underworld*. Scribner, 1997.
———. *White Noise*. Viking, 1985.
DeRosa, Aaron. "The End of Futurity: Proleptic Nostalgia and the War on Terror." *Literature Interpretation Theory*, vol. 25, 2014, pp. 88–107.
Dick, Philip K. *Vulcan's Hammer*. 1960. Vintage Books, 2004.
Dimock, Wai Chee. "Introduction: Planet and America, Set and Subset." *Shades of the Planet: American Literature as World Literature*, Princeton University Press, 2007, pp. 1–16.
Dinnen, Zara. *The Digital Banal: New Media and American Literature and Culture*. Columbia University Press, 2018.
Dostoevsky, Fyodor. "Poor Folk." *Poor Folk and Other Stories*. Translated by David McDuff, Penguin Books, 1988, pp. 1–130.
Dreyfus, Hubert L. "Anonymity versus Commitment: The Dangers of Education on the Internet." *Philosophy of Technology: The Technological Condition: An Anthology*, 2nd ed., edited by Robert C. Scharff and Val Dusek, Wiley Blackwell, 2014, pp. 641–647.
Dubay, Eric. "200 Proofs Earth Is Not a Spinning Ball." YouTube, uploaded by Planet Plane, 15 Sept. 2016, https://youtu.be/-Ax_YpQsy88. Accessed 15 July 2019.
Dusek, Val. *Philosophy of Technology: An Introduction*. Blackwell, 2006.

Edwards, Brian T., and Dilip Parameshwar Gaonkar. "Introduction: Globalizing American Studies." *Globalizing American Studies*. University of Chicago Press, 2010, pp. 1–44.
Egan, Jennifer. "Black Box." *New Yorker*, 28 May 2012, https://www.newyorker.com/magazine/2012/06/04/black-box-2. Accessed 4 June 2019.
———. *Look at Me*. Anchor Books, 2001.
———. "Rewiring the Real." Interview with Willing Davidson. Institute for Religion, Culture, and Public Life, Columbia University, 7 Feb. 2012.
———. *A Visit from the Goon Squad*. Anchor Books, 2011.
Eggers, Dave. *The Circle*. Vintage Books, 2013.
———. "Dave Eggers." *Monocle*, 4 Mar. 2018, https://monocle.com/radio/shows/the-big-interview/75. Accessed 11 Feb. 2019.
———. "Dave Eggers: 'I always picture Trump hiding under a table.'" Interview with Paul Laity. *The Guardian*, 22 June 2018, https://www.theguardian.com/books/2018/jun/22/dave-eggers-interview-circle-lifters. Accessed 11 Feb. 2019.
———. *A Hologram for the King*. Vintage Books, 2012.
———. "A Short Interview with Dave Eggers about the New Film, *The Circle*." *McSweeney's Internet Tendency*, 26 Apr. 2017, https://www.mcsweeneys.net/articles/a-short-interview-with-dave-eggers-about-the-new-film-the-circle. Accessed 11 Feb. 2019.
———. *What Is the What: The Autobiography of Valentino Achak Deng*. 2006. Vintage Books, 2007.
———. *Zeitoun*. 2009. Vintage Books, 2010.
Einstein, Albert. "The Special Theory of Relativity." *Relativity: The Special and General Theory*, 100th anniversary ed., edited by Hanoch Gutfreund and Jürgen Renn, Princeton University Press, 2015, pp. 11–71.
Elias, Amy J. "History." *The Cambridge Companion to Thomas Pynchon*, edited by Inger H. Dalsgaard, Luc Herman, and Brian McHale, Cambridge University Press, 2012, pp. 123–135.
Eliot, T. S. "Ash-Wednesday." 1936. *T. S. Eliot: The Collected Poems, 1909–1962*. Harcourt, Brace & World, 1963, pp. 83–96.
Fan, Lai-Tze. "The Digital Intensification of Postmodern Poetics." *The Poetics of Genre in the Contemporary Novel*, edited by Tim Lanzendörfer, Lexington Books, 2016, pp. 35–56.
Fassler, Joe. "The Joyce Carol Oates Story That Shares DNA with 'Cat Person.'" *The Atlantic*, 16 Jan. 2019, https://www.theatlantic.com/entertainment/archive/2019/01/kristen-roupenians-cat-person-inspiration/580498. Accessed 2 Mar. 2019.
Ferris, Joshua. "Always on Display: An Interview with Joshua Ferris." Interview with Jonathan Lee. *Paris Review*, 19 May 2014, https://www.theparisreview.org/blog/2014/05/19/an-interview-with-joshua-ferris. Accessed 25 Apr. 2019.
———. "A Conversation with Joshua Ferris." Reading Group Guide: *Then We Came to the End*, in *Then We Came to the End*, Back Bay Books, 2007, pp. 3–8.
———. "Fragments." *The Dinner Party and Other Stories*, Little, Brown, 2017, pp. 159–172.
———. "Joshua Ferris." Interview with Christopher Bollen. *Interview*, 12 May 2014, https://www.interviewmagazine.com/culture/joshua-ferris. Accessed 25 Apr. 2019.
———. "Joshua Ferris: The Writer on Hard Work, an Obsessive Love of Sports—and Failing to Keep up with Technology." Interview with Darren Richman. *Independent*, 1 June 2014,

https://www.independent.co.uk/arts-entertainment/books/features/joshua-ferris-the-writer-on-hard-work-an-obsessive-love-of-sports-and-failing-to-keep-up-with-9457512.html. Accessed 25 Apr. 2019.

———. "Joshua Ferris: Why Comedy Is 'Utterly Essential.'" Interview with Mary Laura Philpott. *Musing*, 17 Mar. 2015, https://parnassusmusing.net/2015/03/17/joshuaferris. Accessed 25 Apr. 2019.

———. "More Abandon (or Whatever Happened to Joe Pope?)." *The Dinner Party and Other Stories*, Little, Brown, 2017, pp. 135–157.

———. *Then We Came to the End*. Back Bay Books, 2007.

———. *To Rise Again at a Decent Hour*. Back Bay Books, 2014.

———. "The World of Writer Joshua Ferris." Interview with Edie Greaves. *The Telegraph*, 30 May 2014, https://www.telegraph.co.uk/culture/books/authorinterviews/10860378/The-world-of-writer-Joshua-Ferris.html. Accessed 25 Apr. 2019.

Fincher, David, director. *The Social Network*. Columbia Pictures, 2010.

Finkel, Eli J., and Susan Sprecher. "The Scientific Flaws of Online Dating Sites: What the 'Matching Algorithms' Miss." *Scientific American*, 8 May 2012, https://www.scientificamerican.com/article/scientific-flaws-online-dating-sites. Accessed 2 Mar. 2019.

Foer, Jonathan Safran. *Everything Is Illuminated*. Penguin Books, 2002.

———. *Extremely Loud and Incredibly Close*. Houghton Mifflin, 2005.

———. *Here I Am*. Farrar, Straus and Giroux, 2016.

———. "Interview with Jonathan Safran Foer." Interview with Jade Chang. Goodreads, 6 Sept. 2016, https://www.goodreads.com/interviews/show/1170.Jonathan_Safran_Foer. Accessed 30 May 2019.

———. "Jonathan Safran Foer Interview: 'The meaning of the book is not something that is mine.'" Interview with Karen Heller. *Independent*, 8 Sept. 2016, https://www.independent.co.uk/arts-entertainment/books/features/jonathan-safran-foer-interview-the-meaning-of-the-book-is-not-something-that-is-mine-a7231901.html. Accessed 30 May 2019.

———. "Jonathan Safran Foer on Marriage, Religion and Universal Balances." Interview with Terry Gross. NPR, 11 Oct. 2016, https://www.npr.org/2016/10/11/497515290/jonathan-safran-foer-on-marriage-religion-and-universal-balances. Accessed 30 May 2019.

———. "A Primer for the Punctuation of Heart Disease." *New Yorker*, 10 June 2002, https://www.newyorker.com/magazine/2002/06/10/a-primer-for-the-punctuation-of-heart-disease. Accessed 30 May 2019.

———. "Technology Is Diminishing Us." *The Guardian*, 3 Dec. 2016, www.theguardian.com/books/2016/dec/03/jonathan-safran-foer-technology-diminishing-us. Accessed 30 May 2019.

———. *Tree of Codes*. Visual Editions, 2010.

Forster, E. M. *Howards End*. 1910. Penguin Books, 2000.

Foster, Hannah W. *The Coquette*, edited by Cathy N. Davidson, Oxford University Press, 1986.

Franzen, Jonathan. "The Reader in Exile." *How to Be Alone: Essays*, Farrar, Straus and Giroux, 2002, pp. 164–178.

Friedman, Thomas L. "Is Google God?" *New York Times*, 29 June 2003, https://www.nytimes.com/2003/06/29/opinion/is-google-god.html. Accessed 24 June 2019.

———. "Online and Scared." *New York Times*, 11 Jan. 2017, https://www.nytimes.com/2017/01/11/opinion/online-and-scared.html. Accessed 24 June 2019.

———. *The World Is Flat: A Brief History of the Twenty-first Century*. Farrar, Straus and Giroux, 2005.

Fukuyama, Francis. "The End of History?" *National Interest*, Summer 1989, pp. 3–18.

———. *The End of History and the Last Man*. Free Press, 1992.

———. *Our Posthuman Future: Consequences of the Biotechnology Revolution*. Farrar, Straus and Giroux, 2002.

Galow, Timothy W. *Understanding Dave Eggers*. University of South Carolina Press, 2014.

Gamal, Ahmed. "The Global and the Postcolonial in Post-migratory Literature." *Journal of Postcolonial Writing*, vol. 49, no. 5, 2013, pp. 596–608.

Gibson, William. "Burning Chrome." 1982. *Burning Chrome*, Ace Books, 1986.

———. *Neuromancer*. 1984. Ace Books, 2000.

Gleich, Lewis S. "Ethics in the Wake of the Image: The Post-9/11 Fiction of DeLillo, Auster, and Foer." *Journal of Modern Literature*, vol. 37, no. 3, 2014, pp. 161–176.

Goebel, Rolf J. "From Postcoloniality to Global Media Culture: Multimedial Reflections on Metropolitan Space." *Re-Inventing the Postcolonial (in the) Metropolis*, edited by Cecile Sandten and Annika Bauer, Brill Rodopi, 2016, pp. 327–341.

Gordon, Douglas. *24 Hour Psycho*. 1993, Museum of Modern Art, New York.

Grady, Constance. "The Uproar Over the *New Yorker* Short Story 'Cat Person,' Explained." *Vox*, 12 Dec. 2017, https://www.vox.com/culture/2017/12/12/16762062/cat-person-explained-new-yorker-kristen-roupenian-short-story. Accessed 2 Mar. 2019.

Graff, Harvey J. *The Legacies of Literacy: Continuities and Contradictions in Western Culture and Society*. Indiana University Press, 1987.

Griffey, Harriet. "The Lost Art of Concentration: Being Distracted in a Digital World." *The Guardian*, 14 Oct. 2018, https://www.theguardian.com/lifeandstyle/2018/oct/14/the-lost-art-of-concentration-being-distracted-in-a-digital-world. Accessed 2 Mar. 2019.

Gumprecht, Blake. *The American College Town*. University of Massachusetts Press, 2008.

Gurak, Laura J., and Smiljana Antonijevic. "Digital Rhetoric and Public Discourse." *The Sage Handbook of Rhetorical Studies*, edited by Andrea A. Lunsford, Sage, 2009, pp. 497–507.

Haeselin, David. "Welcome to the Indexed World: Thomas Pynchon's *Bleeding Edge* and the Things Search Engines Will Not Find." *Critique: Studies in Contemporary Fiction*, vol. 58, no. 4, 2017, pp. 313–324.

Hamid, Mohsin. *Exit West*. Riverhead Books, 2017.

———. "The (Former) General in His Labyrinth." Penguin, 23 Apr. 2008, wetellstories.co.uk/stories/week6. Accessed 6 December 2008.

———. "The Great Divide." *New York Times Magazine*, 18 Feb. 2015, https://www.nytimes.com/2015/02/22/magazine/the-great-divide.html. Accessed 24 June 2015.

———. *How to Get Filthy Rich in Rising Asia*. Riverhead Books, 2013.

———. "Mohsin Hamid on the Dangers of Nostalgia: We Need to Imagine a Brighter Future."

The Guardian, 25 Feb. 2017. www.theguardian.com/books/2017/feb/25/mohsin-hamid-danger-nostalgia-brighter-future. Accessed 1 Sept. 2017.

———. "Mohsin Hamid on the Rise of Nationalism: 'In the land of the pure, no one is pure enough.'" *The Guardian*, 27 Jan. 2018, https://www.theguardian.com/books/2018/jan/27/mohsin-hamid—exit-west-pen-pakistan. Accessed 24 June 2019.

———. *Moth Smoke*. Picador, 2000.

———. *The Reluctant Fundamentalist*. Houghton Mifflin Harcourt, 2007.

Hammond, Adam. *Literature in the Digital Age: An Introduction*. Cambridge University Press, 2016.

Hansen, Randall. "State Controls: Borders, Refugees, and Citizenship." *The Oxford Handbook of Refugee and Forced Migration Studies*, edited by Elena Fiddian-Qasmiyeh, Gil Loescher, Katy Long, and Nando Sigona, Oxford University Press, 2014, pp. 253–264.

Haraway, Donna. "A Cyborg Manifesto: Science, Technology, and Socialist Feminism in the Late Twentieth Century." *Philosophy of Technology: The Technological Condition: An Anthology*, 2nd ed., edited by Robert C. Scharff and Val Dusek, Wiley Blackwell, 2014, pp. 610–630.

Hayles, N. Katherine. *Electronic Literature: New Horizons for the Literary*. University of Notre Dame Press, 2008.

———. "The Future of Literature: Complex Surfaces of Electronic Texts and Print Books." *A Time for the Humanities: Futurity and the Limits of Autonomy*, edited by James J. Bono, Tim Dean, and Ewa Plonowska Ziarek, Fordham University Press, 2008, pp. 180–209.

———. *How We Think: Digital Media and Contemporary Technogenesis*. University of Chicago Press, 2012.

Heidegger, Martin. "The Question Concerning Technology." *Philosophy of Technology: The Technological Condition: An Anthology*, 2nd ed., edited by Robert C. Scharff and Val Dusek, Wiley Blackwell, 2014, pp. 305–317.

Heschel, Abraham Joshua. *Man Is Not Alone: A Philosophy of Religion*. Farrar, Straus and Giroux, 1951.

Huehls, Mitchum. "The Great Flattening." *Contemporary Literature*, vol. 54, no. 4, 2013, pp. 861–871.

Humann, Heather Duerre. "Nachträglichkeit and 'Narrative Time' in Jennifer Egan's *A Visit from the Goon Squad*." *Pennsylvania Literary Journal*, vol. 9, no. 2, 2017, pp. 85–97.

Hungerford, Amy. "Don DeLillo's Latin Mass." *Contemporary Literature*, vol. 47, no. 3, 2006, pp. 343–380.

Hutcheon, Linda. *A Poetics of Postmodernism: History, Theory, and Fiction*. Routledge, 1988.

Hutton, Margaret-Anne. "Plato, New Media Technologies, and the Contemporary Novel." *Mosaic*, vol. 51, no. 1, 2018, pp. 179–195.

Hwang, Jung-Suk. "Staging the Uneven World of Cybercapitalism on 47th Street in Don DeLillo's *Cosmopolis*." *Critique: Studies in Contemporary Fiction*, vol. 59, no. 1, 2018, pp. 27–40.

Ihde, Don. *Bodies in Technology*. University of Minnesota Press, 2002.

———. *Technology and the Lifeworld: From Garden to Earth*. Indiana University Press, 1990.

Jackson, Shelley. *Patchwork Girl, or, a Modern Monster*. Eastgate Systems, 1995.

Jameson, Fredric. *The Political Unconscious: Narrative as a Socially Symbolic Act*. Cornell University Press, 1981.

———. *Postmodernism, or, the Cultural Logic of Late Capitalism*. Duke University Press, 1991.

Jeffrey, David Lyle. "Meditation and Atonement in the Art of Marc Chagall." *Religion and the Arts*, vol. 16, no. 3, 2012, pp. 211–230.

Johnston, John. *Information Multiplicity: American Fiction in the Age of Media Saturation*. Johns Hopkins University Press, 1998.

Joyce, Michael. *afternoon, a story*. Eastgate Systems, 1990.

Junod, Tom. "The Man Who Invented 9/11." *Esquire*, 7 May 2007, https://www.esquire.com/entertainment/books/reviews/a2942/delillo. Accessed 19 Dec. 2018.

Kermode, Frank. *The Sense of an Ending*. 1966. Oxford University Press, 2000.

Keyes, Ralph. *The Post-Truth Era: Dishonesty and Deception in Contemporary Life*. St. Martin's Press, 2004.

Khazan, Olga. "A Viral Short Story for the #MeToo Moment." *The Atlantic*, 11 Dec. 2017, https://www.theatlantic.com/technology/archive/2017/12/a-viral-short-story-for-the-metoo-moment/548009. Accessed 2 Mar. 2019.

Kierkegaard, Søren. *The Present Age: On the Death of Rebellion*. Translated by Alexander Dru, Harper Perennial, 2010.

Kingsley, Patrick. *The New Odyssey: The Story of Europe's Refugee Crisis*. Guardian Books and Faber & Faber, 2016.

Kingsolver, Barbara. *Flight Behavior*. Harper Perennial, 2012.

Kruger, Anique. "'A fervid intensity of connectedness': Zadie Smith, the Cosmopolitan Novel, and the Ethics of Community." *Oxford Research in English*, vol. 2, 2015, pp. 68–84.

Kubrick, Stanley, director. *2001: A Space Odyssey*. MGM, 1968.

Kundera, Milan. *The Unbearable Lightness of Being*. Translated by Michael Henry Heim, Harper & Row, 1984.

Kwan-Terry, John. "*Ash-Wednesday*: A Poetry of Verification." *The Cambridge Companion to T. S. Eliot*, edited by A. David Moody, Cambridge University Press, 1994, pp. 132–141.

Laist, Randy. *Technology and Postmodern Subjectivity in Don DeLillo's Novels*. Peter Lang Publishing, 2010.

Landrum, Ashley, A. Olshansky, O. Richards, and T. Plate. "YouTube as a Primary Propagator of Flat Earth Philosophy." YouTube: Friend or Foe to Communicating about Science and Health, Annual Meeting of the American Association for the Advancement of Science, Feb. 2019, Washington, D.C.

Lanier, Jaron. *You Are Not a Gadget: A Manifesto*. Vintage Books, 2010.

Lankshear, Colin, and Peter L. McLaren. Introduction. *Critical Literacy: Politics, Praxis, and the Postmodern*. State University of New York Press, 1993, pp. 1–56.

Lechner, Frank J., and John Boli, editors. *The Globalization Reader*. 4th ed., Wiley-Blackwell, 2012.

LeClair, Tom. *In the Loop: Don DeLillo and the Systems Novel*. University of Illinois Press, 1987.

Leps, Marie-Christine. "How to Map the Non-Place of Empire: DeLillo's *Cosmopolis*." *Textual Practice*, vol. 28, no. 2, 2014, pp. 305–327.

Lin, Tao. *Shoplifting from American Apparel*. Melville House, 2009.

Loh, Lucienne. "Zadie Smith's Short Stories: Englishness in a Globalized World." *Reading Zadie Smith: The First Decade and Beyond*, edited by Philip Tew, Bloomsbury, 2014, pp. 169–185.

Luce, Henry. "The American Century." *Life*, 17 Feb. 1941, pp. 61–65.

Lynch, Michael Patrick. *The Internet of Us: Knowing More and Understanding Less in the Age of Big Data*. Liveright, 2016.

Mączyńska, Magdalena. "'That God Chip in the Brain': Religion in the Fiction of Zadie Smith." *Reading Zadie Smith: The First Decade and Beyond*, edited by Philip Tew, Bloomsbury, 2014, pp. 127–139.

Malchik, Antonia. "The Problem with Social-Media Protests." *The Atlantic*, 6 May 2019, https://www.theatlantic.com/technology/archive/2019/05/in-person-protests-stronger-online-activism-a-walking-life/578905. Accessed 5 Dec. 2019.

Malewitz, Raymond. "'Some new dimension devoid of hip and bone': Remediated Bodies and Digital Posthumanism in Gary Shteyngart's *Super Sad True Love Story*." *Arizona Quarterly: A Journal of American Literature, Culture, and Theory*, vol. 71, no. 4, 2015, pp. 107–127.

Manjoo, Farhad. "Welcome to the Post-Text Future." *New York Times*, 14 Feb. 2018, https://www.nytimes.com/interactive/2018/02/09/technology/the-rise-of-a-visual-internet.html. Accessed 24 June 2019.

Marx, Karl, and Friedrich Engels. "Manifesto of the Communist Party." *The Marx-Engels Reader*, 2nd ed., edited by Robert C. Tucker, W. W. Norton, 1978, pp. 469–500.

Mather, Cotton. "The Wonders of the Invisible World." John Russell Smith, 1862. https://www.gutenberg.org/files/28513/28513-h/28513-h.htm.

The Matrix. Directed by the Wachowski Brothers, Warner Brothers, 1999.

The Matrix Reloaded. Directed by the Wachowski Brothers, Warner Brothers, 2003.

The Matrix Revolutions. Directed by the Wachowski Brothers, Warner Brothers, 2003.

McCann, Colum. *Let the Great World Spin*. Random House, 2009.

McChesney, Robert W. *Digital Disconnect: How Capitalism Is Turning the Internet against Democracy*. New Press, 2013.

McClure, John A. "DeLillo and Mystery." *The Cambridge Companion to Don DeLillo*, edited by John N. Duvall, Cambridge University Press, 2008, pp. 166–178.

McGrath, Alister, and Joanna Collicutt McGrath. *The Dawkins Delusion? Atheist Fundamentalism and the Denial of the Divine*. InterVarsity Press, 2007.

Melville, Herman. "Bartleby, the Scrivener: A Story of Wall-Street." 1853. *The Norton Anthology of American Literature: American Literature 1820–1865*, 6th ed., edited by Nina Baym, W. W. Norton, 2003, vol. B, pp. 2330–2355.

———. *Moby-Dick, or, The Whale*. 1851. Penguin Books, 2003.

Miller, Vincent. *Understanding Digital Culture*. SAGE, 2011.

Misa, Thomas J. *Leonardo to the Internet: Technology and Culture from the Renaissance to the Present*. 2nd ed., Johns Hopkins University Press, 2011.

Moling, Martin. "'No Future': Time, Punk Rock and Jennifer Egan's *A Visit from the Goon Squad*." *Arizona Quarterly: A Journal of American Literature, Culture, and Theory*, vol. 72, no. 1, 2016, pp. 51–77.

Morey, Peter. "'The rules of the game have changed': Mohsin Hamid's *The Reluctant Fundamentalist* and Post-9/11 Fiction." *Journal of Postcolonial Writing*, vol. 47, no. 2, 2011, pp. 135–146.

Morgan, Katrina. "Which Sounds Better, Analog or Digital Music?" *Scientific American*, 11 Oct. 2017, https://blogs.scientificamerican.com/observations/which-sounds-better-analog-or-digital-music. Accessed 4 June 2019.

Must Love Dogs. Directed by Gary David Goldberg, Warner Brothers, 2005.

Nabokov, Vladimir. *Lolita*. 1955. Vintage International, 1997.

Nealon, Jeffrey T. *Post-Postmodernism, or, the Logic of Just-in-Time Capitalism*. Stanford University Press, 2012.

The Net. Directed by Irwin Winkler, Columbia Pictures, 1995.

The New London Group. "A Pedagogy of Multiliteracies: Designing Social Futures." *Harvard Educational Review*, vol. 66, no. 1, 1996, pp. 60–92.

Noble, David F. *The Religion of Technology: The Divinity of Man and the Spirit of Invention*. Alfred A. Knopf, 1997.

Olster, Stacey. *The Cambridge Introduction to Contemporary American Fiction*. Cambridge University Press, 2017.

Pannapacker, William. "The MLA and the Digital Humanities." *Chronicle of Higher Education*, 28 Dec. 2009, https://www.chronicle.com/blogPost/The-MLA-the-Digital/19468. Accessed 19 June 2019 through the Wayback Machine.

Pariser, Eli. *The Filter Bubble: How the Personalized Web Is Changing What We Read and How We Think*. Penguin Press, 2011.

Pew Research Center. "Political Polarization in the American Public." Pew Research Center, 12 June 2014, https://www.people-press.org/2014/06/12/political-polarization-in-the-american-public. Accessed 10 Dec. 2019.

Plato. *The Republic*. 2nd ed., translated by Desmond Lee, Penguin Books, 1974.

Poon, Angelia. "Helping the Novel: Neoliberalism, Self-Help, and the Narrating of the Self in Mohsin Hamid's *How to Get Filthy Rich in Rising Asia*." *Journal of Commonwealth Literature*, vol. 52, no. 1, 2017, pp. 139–150.

Postman, Neil. *Technopoly: The Surrender of Culture to Technology*. Knopf, 1992.

Powers, Richard. *Galatea 2.2*. Picador, 1995.

Prabhu, Anjali. *Hybridity: Limits, Transformations, Prospects*. State University of New York Press, 2007.

Pressman, Jessica. *Digital Modernism: Making It New in New Media*. Oxford University Press, 2014.

Pursell, Carroll. *Technology in Postwar America: A History*. Columbia University Press, 2007.

Pynchon, Thomas. *Against the Day*. Penguin Books, 2006.

———. *Bleeding Edge*. Penguin Press, 2013.

———. *The Crying of Lot 49*. 1965. Perennial Classics, 1999.

———. "Entropy." *Slow Learner: Early Stories*. Little, Brown, 1984, pp. 79–98.

———. *Gravity's Rainbow*. Penguin Books, 1973.

———. *Inherent Vice*. Penguin Press, 2009.

———. "Is It O.K. to Be a Luddite?" *New York Times*, 28 Oct. 1984, https://archive.nytimes

.com/www.nytimes.com/books/97/05/18/reviews/pynchon-luddite.html. Accessed 3 June 2019.

Rothko, Mark. "The Romantics Were Prompted." 1947. *Writings on Art*, edited by Miguel López-Remiro, Yale University Press, 2006, pp. 58–59.

———. Rothko Chapel. 1972, Houston.

Roupenian, Kristen. "The Boy in the Pool." *You Know You Want This: "Cat Person" and Other Stories*, Scout Press, 2019, pp. 149–168.

———. "Cat Person." *You Know You Want This: "Cat Person" and Other Stories*, Scout Press, 2019, pp. 77–98.

———. "Death Wish." *You Know You Want This: "Cat Person" and Other Stories*, Scout Press, 2019, pp. 201–214.

———. "The Good Guy." *You Know You Want This: "Cat Person" and Other Stories*, Scout Press, 2019, pp. 99–148.

———. "Kristen Roupenian on the Self-Deceptions of Dating." Interview with Deborah Treisman. *New Yorker*, 4 Dec. 2017, https://www.newyorker.com/books/this-week-in-fiction/fiction-this-week-kristen-roupenian-2017-12-11. Accessed 2 Mar. 2019.

———. "Look at Your Game, Girl." *You Know You Want This: "Cat Person" and Other Stories*, Scout Press, 2019, pp. 13–26.

———. "Scarred." *You Know You Want This: "Cat Person" and Other Stories*, Scout Press, 2019, pp. 169–180.

———. "What It Felt Like When 'Cat Person' Went Viral." *New Yorker*, 10 Jan. 2019, https://www.newyorker.com/books/page-turner/what-it-felt-like-when-cat-person-went-viral. Accessed 2 Mar. 2019.

———. "With *You Know You Want This*, Kristen Roupenian Proves She's Much More Than a 'Cat Person.'" Interview with David Canfield. *Entertainment Weekly*, 10 Jan. 2019, https://ew.com/author-interviews/2019/01/10/kristen-roupenian-cat-person-profile. Accessed 2 Mar. 2019.

Rushdie, Salman. "Angel Gabriel." *London Review of Books*, vol. 4, no. 17, 16 Sept. 1982, https://www.lrb.co.uk/v04/n17/salman-rushdie/angel-gabriel. Accessed 23 Mar. 2018.

———. *East, West: Stories*. Vintage Books, 1994.

———. *The Satanic Verses*. 1988. Random House, 2008.

Russell, Alison. "The One and the Many: Joshua Ferris's *Then We Came to the End*." *Critique: Studies in Contemporary Fiction*, vol. 59, no. 3, 2018, pp. 319–331.

Ruthven, Malise. *Fundamentalism: A Very Short Introduction*. Oxford University Press, 2007.

Sample, Ian. "Study Blames YouTube for Rise in Number of Flat Earthers." *The Guardian*, 17 Feb. 2019, https://www.theguardian.com/science/2019/feb/17/study-blames-youtube-for-rise-in-number-of-flat-earthers. Accessed 25 Apr. 2019.

Schmidt, Eric, and Jared Cohen. *The New Digital Age: Transforming Nations, Businesses, and Our Lives*. Vintage Books, 2013.

Schradie, Jen. *The Revolution That Wasn't: How Digital Activism Favors Conservatives*. Harvard University Press, 2019.

Schreibman, Susan, Ray Siemens, and John Unsworth. Preface. *A New Companion to Digital Humanities*, by Schreibman, Siemens, and Unsworth, John Wiley & Sons, 2016.

Schulz, Bruno. *The Street of Crocodiles*. 1933. Penguin Press, 1992.

Sciolino, Martina. "The Contemporary American Novel as World Literature: The Neoliberal Antihero in Don DeLillo's *Cosmopolis*." *Texas Studies in Literature and Language*, vol. 57, no. 2, 2015, pp. 210–241.
Seed, David. *Science Fiction: A Very Short Introduction*. Oxford University Press, 2011.
Shelley, Mary. *Frankenstein*. 1818. 2nd ed., edited by J. Paul Hunter, W. W. Norton, 2012.
Shirky, Clay. *Cognitive Surplus*. Penguin Press, 2010.
Shteyngart, Gary. "A Conversation with Gary Shteyngart." Interview with Elizabeth Tannen. *Blue Mesa Review*, vol. 24, 2011, pp. 170–178.
———. *Little Failure: A Memoir*. Random House, 2014.
———. "O.K., Glass: Confessions of a Google Glass Explorer." *New Yorker*, 29 July 2013, https://www.newyorker.com/magazine/2013/08/05/o-k-glass. Accessed 4 June 2019.
———. *The Russian Debutante's Handbook*. Riverhead Books, 2002.
———. *Super Sad True Love Story*. Random House, 2010.
Smith, Roberta. "Art; The World According to Judd." *New York Times*, 26 Feb. 1995, https://www.nytimes.com/1995/02/26/arts/art-the-world-according-to-judd.html. Accessed 11 Feb. 2019.
Smith, Zadie. "Crazy They Call Me." *New Yorker*, 27 Feb. 2017, www.newyorker.com/magazine/2017/03/06/crazy-they-call-me. Accessed 9 Dec. 2017.
———. "Elegy for a Country's Seasons." *New York Review of Books*, 3 Apr. 2014, www.nybooks.com/articles/2014/04/03/elegy-countrys-seasons. Accessed 5 Dec. 2017.
———. "Escape from New York." *New Yorker*, 1 June 2015, https://www.newyorker.com/magazine/2015/06/08/escape-from-new-york. Accessed 9 Dec. 2017.
———. "Generation Why?" *New York Review of Books*, 25 Nov. 2010, www.nybooks.com/articles/2010/11/25/generation-why. Accessed 5 Dec. 2017.
———. "Meet the President!" *New Yorker*, 5 Aug. 2013, www.newyorker.com/magazine/2013/08/12/meet-the-president. Accessed 9 Dec. 2017.
———. *NW*. Penguin Books, 2012.
———. *On Beauty*. Penguin Books, 2005.
———. *White Teeth*. Vintage International, 2000.
———. "Zadie, Take Three." Interview with Jessica Murphy Moo. *The Atlantic*, Oct. 2005, https://www.theatlantic.com/magazine/archive/2005/10/zadie-take-three/304294. Accessed 11 Dec. 2017.
Starre, Alexander. *Metamedia: American Book Fictions and Literary Print Culture after Digitization*. University of Iowa Press, 2015.
Steger, Manfred B. *Globalization: A Very Short Introduction*. Oxford University Press, 2013.
Stephenson, Neal. *Snow Crash*. Bantam Books, 1992.
Sunstein, Cass R. *Infotopia: How Many Minds Produce Knowledge*. Oxford University Press, 2006.
Tales from the Thousand and One Nights. Translated by N. J. Dawood, Penguin Books, 1973.
Tanakh: A New Translation of the Holy Scriptures According to the Traditional Hebrew Text. Jewish Publication Society, 1985.
Tolstoy, Leo. *War and Peace*. 1869. Translated by Richard Pevear and Larissa Volokhonsky, Alfred A. Knopf, 2007.
Trendel, Aristi. "The Posthuman in Don DeLillo's *Cosmopolis*." *War on the Human: New Re-

sponses to an Ever-Present Debate, edited by Theodora Tsimpouki and Konstantinos Blatanis, Cambridge Scholars Publishing, 2017, pp. 115–129.

The Truth about Cats and Dogs. Directed by Michael Lehmann, Twentieth Century Fox, 1996.

Turkle, Sherry. *Alone Together: Why We Expect More from Technology and Less from Each Other*. Basic Books, 2011.

———. *Life on the Screen: Identity in the Age of the Internet*. Simon & Schuster, 1995.

———. *Reclaiming Conversation: The Power of Talk in a Digital Age*. Penguin Books, 2015.

———. *The Second Self: Computers and the Human Spirit*. 1984. Twentieth anniversary edition, MIT Press, 2005.

Updike, John. *Roger's Version*. Fawcett Columbine, 1986.

Vanderwees, Chris. "Photographs of Falling Bodies and the Ethics of Vulnerability in Jonathan Safran Foer's *Extremely Loud and Incredibly Close*." *Canadian Review of American Studies*, vol. 45, no. 2, 2015, pp. 174–194.

Varsava, Jerry A. "The 'Saturated Self': Don DeLillo on the Problem of Rogue Capitalism." *Contemporary Literature*, vol. 46, no. 1, 2005, pp. 78–107.

Vonnegut, Kurt. *Player Piano*. Delacorte Press, 1952.

Wajcman, Judy. *Pressed for Time: The Acceleration of Life in Digital Capitalism*. University of Chicago Press, 2015.

Wallace, David Foster. "E Unibus Pluram: Television and U.S. Fiction." *Review of Contemporary Fiction*, vol. 13, no. 2, 1993, pp. 151–194.

———. "My Appearance." *Girl with Curious Hair*. W. W. Norton, 1989, pp. 175–201.

Wanner, Adrian. "Russian Hybrids: Identity in the Translingual Writings of Andreï Makine, Wladimir Kaminer, and Gary Shteyngart." *Slavic Review*, vol. 67, no. 3, 2008, pp. 662–681.

Weber, Max. *The Protestant Ethic and the "Spirit" of Capitalism and Other Writings*. 1905, edited by Peter Baehr and Gordon C. Wells, Penguin Books, 2002.

Winthrop, John. "A Model of Christian Charity." *The Norton Anthology of American Literature: Literature to 1820*, 6th ed., edited by Nina Baym, Norton, 2003, vol. A, pp. 206–217.

Wurth, Kiene Brillenburg. "Old and New Medialities in Foer's *Tree of Codes*." *CLCWeb: Comparative Literature and Culture*, vol. 13, no. 3, 2011, n.p.

Wysocki, Anne Frances. "with eyes that think, and compose, and think: ON VISUAL RHETORIC." *Teaching Writing with Computers: An Introduction*, edited by Pamela Takayoshi and Brian Huot, Houghton Mifflin, 2003, pp. 182–201.

INDEX

activism, 32, 144, 184; online, 6–7, 189n5
Adams, Henry, 9, 46
Afghanistan, 167
afterlife, 57, 96, 127, 161
Against the Day (Pynchon), 55
Alang, Navneet, 195n1
Alexie, Sherman, 6
allegory, 11, 24, 140, 142
Allende, Salvador, 54
Allington, Daniel, 5
Alone Together (Turkle), 19, 97, 123, 150, 165, 166
al-Qaeda, 65
American Century, 159, 184–187; Egan and, 75; Eggers on, 114; Luce on, 1, 51, 152, 184, 185; U.S. exceptionalism and, 16, 75, 147, 184–185
American Dream, 3
American Studies, 146–147, 185
analog versus digital technology, 49–50, 64–65, 70–71
Anderson, Craig A., 99
Anderson, M. T., 13
Antonijevic, Smiljana, 13, 89
apocalypse, 30, 55, 59; Cold War, 51–53; cybercapitalism and, 138–139; Manjoo on, 180; Zadie Smith on, 194n13
Arab Spring, 113
Armstrong, Karen, 82

ARPANET, 6, 53, 157
Asimov, Isaac, 11, 24
Assange, Julian, 6, 86
Atchison, S. Todd, 191n9
Attebery, Brian, 190n10
Auster, Paul, 69

"badasses" (Pynchon), 48, 50, 61–62, 64, 75, 191n3
Ball, Cheryl E., 183
Barth, John, 13, 189n2
Barth, Karl, 191n5
Baudrillard, Jean, 95
Bauman, Zygmunt, 150, 165, 194n6, 195n16
Beaumont, Matt, 13, 30
Beckert, Sven, 115
Beckett, Andy, 185
Bergholtz, Benjamin, 148, 193n3
Best, Stephen, 182
Bewes, Timothy, 182
Bhabha, Homi K., 19, 46, 77, 183–184; hybridity defined by, 8; hybridization of capitalism and, 112; Roupenian and, 70; Zadie Smith and, 147; Turkle and, 149–150
Bible, 86–87; Genesis, 80, 103–105, 107, 109; Job, 87; Revelation, 30, 55, 59; Sermon on the Mount, 80–81, 152

| 211

bildungsroman, 158, 162
Bilge, Sirma, 184, 195n2
binary code, 8, 14; Egan on, 50, 68; Ferris on, 85; Pynchon on, 47; Roupenian on, 45, 70
Birkerts, Sven, 103, 145, 180, 189n2, 192n10
"Black Box" (Egan), 13, 69, 75, 183
Bleeding Edge (Pynchon), 13, 15–16, 47–64; Egan and, 65, 66, 73–75; Hamid and, 164; Zadie Smith and, 163–178
Blum, Andrew, 56
Boli, John, 76, 152
Boxall, Peter, 12
"Boy in the Pool, The" (Roupenian), 22
Brâncuși, Constantin, 142–143
Bremmer, Ian, 4, 146
Breslin, James E. B., 118
Bridle, James, 5
B-side, 70–71, 75
Buchanan, Brad, 194n8
Burke, Peter, 8, 14
Burke, Tarana, 6
Butler, Octavia, 24

Caidin, Martin, 11
Calegari, Nínive Clements, 192n3
capitalism, 121; Beckert on, 115; Degler on, 115; DeLillo on, 112–113; gender politics of, 62–63; hybridization of, 112; Pynchon on, 57; Ruthven on, 114–115; terrorism and, 53–54, 65–66. *See also* cybercapitalism
Carr, Nicolas, 180
"catfishing," 35
"Cat Person" (Roupenian), 13, 15, 20–23, 34–45, 134, 169
Chagall, Marc, 92
Chambers, Claire, 194n5
Charlton, Colin, 183
Chekhov, Anton, 28
Chilean coup (1973), 53–54
Christianity, 81; apocalypse, 55, 59; Bible, 86–87; Latin Mass, 193n16; Protestant work ethic, 92
Circle, The (Eggers), 13, 16–17, 112–115, 130–145; Ferris and, 81, 89; Foer and, 132; Pynchon and, 54, 55
civil rights movement, 7
Clark, Andy, 7–8; *Natural-Born Cyborgs*, 42, 69, 116, 136, 149, 183
Cloud, Dana L., 181

Codde, Philippe, 109, 192n14
Cohen, Joshua, 13, 56
Cohen, Thierry, 150, 174–175
Cold War, 6; nostalgia for, 51–53; science fiction of, 10–12
Collado-Rodríguez, Francisco, 50
Collins, Patricia Hill, 184, 195n2
commodification of culture, 69, 113, 115, 140
Coogan, Michael, 103
Cosmopolis (DeLillo), 13, 16–17, 111–130, 182; Egan and, 66, 73, 121; Ferris and, 119, 126; Pynchon and, 54, 73
Covid-19 pandemic, 2, 187
Cowart, David, 47, 49, 52, 116, 190n2
"Crazy They Call Me" (Zadie Smith), 194n4
Crying of Lot 49, The (Pynchon), 47–48, 50, 54, 58, 63
cybercapitalism, 16–17, 111–115; apocalyptic, 138–139; fundamentalism and, 17, 114–115, 132–142, 144; Leps on, 122; Zadie Smith on, 151. *See also* neoliberalism
cyberspace: DeLillo on, 127; Friedman on, 19, 115; Gibson on, 11; Huehls on, 55; "meatspace" versus, 48, 56; Pynchon on, 56–57
"Cyborg Manifesto, A" (Haraway), 7, 19, 23, 43, 46, 183
cyborgs, 151; Caidin on, 11; Clark on, 7–8, 42, 69, 116, 136, 183; DeLillo and, 116; Haraway on, 7, 19, 23, 43, 46, 183; robots, 8, 11, 136, 166

Da, Nan Z., 5–6
Dalsgaard, Inger H., 48
Danielewski, Mark Z., 31
Darlington, Joseph, 190n3
DARPAnet. *See* ARPANET
dating. *See* online dating
Davidson, Willing, 48
Dawkins, Richard, 191n3
"Death Wish" (Roupenian), 22
Degler, Carl, 115
DeLillo, Don, 127, 144, 145, 179–180; "In the Ruins of the Future," 113, 116, 125; *Mao II*, 112; *Point Omega*, 13; "The Power of History," 129; *The Silence*, 13, 112, 122; *Underworld*, 7, 12, 112, 124–125, 127, 192n1; *White Noise*, 112. See also *Cosmopolis*
"depthlessness" (Jameson), 10, 113, 180
DeRosa, Aaron, 49, 69

Desan, Christine, 115
Díaz, Junot, 6
Dick, Philip K., 11
digital art, 64, 150, 174–178
digital culture, 14; DeLillo on, 112–113; Egan on, 49, 67, 70; Franzen on, 171; Hamid on, 174; Pynchon on, 51, 60–61; Shteyngart on, 26, 31; Zadie Smith on, 193n1
digital divide, 150, 164, 166, 170, 194n7
digital humanities, 3–6, 18, 171, 183
"digital rhetoric," 13
digital technology: analog versus, 49–50, 64–65, 70–71; gender politics of, 62–63; religious aura of, 114, 151–152
digitization, 111, 185; art about, 150; DeLillo on, 112, 127, 128; Egan on, 175; Eggers on, 114
Dinnen, Zara, 13–14
Dostoevsky, Fyodor, 30
dot-com bust, 1, 50, 58, 113, 116
Dreyfus, Hubert L., 8
Dubay, Eric, 3
DuPont Corporation, 4, 116
Dusek, Val, 189n7

Eckert, Presper, 189n5
Edwards, Brian T., 146–147, 185
Egan, Jennifer, 47–50, 184; DeRosa on, 69; Ferris and, 81–82; on 9/11 attacks, 48, 65, 69–72
—works of: "Black Box," 13, 69, 75, 183; *Look at Me*, 13, 49; "Rewiring the Real," 48, 70, 73. See also *Visit from the Goon Squad, A*
Eggers, Dave, 179–180; on cybercapitalist fundamentalism, 114–115, 132–142, 144; DeLillo and, 144, 145; Friedman and, 115; on human rights, 114, 133, 139, 144–145, 192n3
—works of: *A Hologram for the King*, 13, 114; *What Is the What*, 114; *Zeitoun*, 114. See also *Circle, The*
Einstein, Albert, 73, 122, 125
electronic literature, 11–12
Electronic Literature (Hayles), 11–12
"Elegy for a Country's Seasons" (Zadie Smith), 152, 194n13
11 September 1973 (Pinochet coup), 53–54. See also 9/11 attacks
Elias, Amy J., 47–48
Eliot, T. S., 161, 163
Emerson, Ralph Waldo, 78

Engels, Friedrich, 120–121
"Entropy" (Pynchon), 52
Everything Is Illuminated (Foer), 79, 102, 192n12
Exit West (Hamid), 13, 17–18, 60, 163–178; DeLillo and, 118; Ferris and, 83; Foer and, 101, 103–104
Extremely Loud and Incredibly Close (Foer), 79, 97, 98, 191nn9–10

Facebook. *See* social media
FaceMash, 148
Farrow, Ronan, 22
Fassler, Joe, 22–23
feminism, 6–7, 22; Haraway on, 7, 43; Pynchon on, 16, 48, 62–63, 186; Roupenian on, 45. *See also* gender politics
Ferris, Joshua, 179, 184; Egan and, 81–82; Eggers and, 81; on identity theft, 57, 181; on print culture, 85; on social media, 77–78
—works of: "Fragments," 89; "More Abandon (Or Whatever Happened to Joe Pope?)," 82; *Then We Came to the End*, 77, 78, 82. See also *To Rise Again at a Decent Hour*
Fiedler, Leslie, 189n2
Fincher, David, 10, 148
Finkel, Eli J., 36
flash poems, 11
flat earthers, 3
flat-world fiction, 18; definition of, 13; science fiction and, 10–18
Foer, Jonathan Safran, 179; on faith, 96; Gross's interview of, 96, 102, 104, 106–109, 192n13; Hamid and, 101, 103–104
—works of: *Everything Is Illuminated*, 79, 102, 192n12; *Extremely Loud and Incredibly Close*, 79, 97, 98, 191nn9–10; "A Primer for the Punctuation of Heart Disease," 79; "Technology Is Diminishing Us," 95, 110; *Tree of Codes*, 79, 80, 110. See also *Here I am*
"(Former) General in His Labyrinth, The" (Hamid), 171–172
Forster, E. M., 148, 193n1
Foster, Hannah W., 30
"Fragments" (Ferris), 89
Franzen, Jonathan, 171, 189n2, 193n10
Friedman, Thomas L., 85–86, 111; on cyberspace, 19, 115; "Is Google God?," 76, 77. See also *World Is Flat, The*
Friends (TV show), 51

Fukuyama, Francis, 51, 159, 194n11
fundamentalism, 76–79, 88, 191n4, 193n3; Armstrong on, 82; Bridle on, 5; cybercapitalist, 17, 114–115, 132–142, 144; Ferris on, 17, 88, 90, 92; Hamid on, 164, 177, 178; terrorism and, 60

Galow, Timothy W., 114
Gaonkar, Dilip Parameshwar, 146–147, 185
García Márquez, Gabriel, 195n15
gender blending, 12, 157
gender politics, 23, 40–42, 45, 59–63; in Hamid's *Exit West*, 168, 169; intersectionality of, 184; literacies of, 15. *See also* feminism
"Generation Why?" (Zadie Smith), 10, 73
gentrification, 56, 58, 172–173
Gibson, William, 11, 24
Giuliani, Rudy, 58
Gleich, Lewis S., 98
globalism, 4, 114–115, 146, 148
globalization: DeLillo on, 112–113; Eggers on, 114; Hamid on, 164, 194n14; literature and, 147; religion and, 76, 152; Shteyngart on, 23, 31; Zadie Smith on, 149–163; Steger on, 146. *See also* neoliberalism
"Globalization 3.0" (Friedman), 1–3, 46, 111, 179
global warming, 152
"Good Guy, The" (Roupenian), 40
Google: Egan on, 74; Friedman on, 76–77; Pariser on, 195n4
Gordon, Douglas, 13
Grady, Constance, 45
Graff, Harvey J., 180–181
Gravity's Rainbow (Pynchon), 47
Gray, Kevin, 112
"Great Divide, The" (Hamid), 187
Griffey, Harriet, 190n5
Ground Zero, 53, 69–70
Gumprecht, Blake, 39
Gurak, Laura J., 13, 89
Gutenberg, Johannes, 84–85, 180

Habitat for Humanity, 93
Haeselin, David, 54
Hamid, Mohsin, 147, 149–150, 180, 184–185; "The (Former) General in His Labyrinth," 171–172; "The Great Divide," 187; *How to Get Filthy Rich in Rising Asia*, 164, 194n14; *Moth Smoke*, 164; *The Reluctant Fundamentalist*, 164, 194n5. See also *Exit West*

Hammond, Adam, 14
Haraway, Donna, 77, 112, 184; "A Cyborg Manifesto," 7, 19, 23, 43, 46, 183
Hayles, N. Katherine, 102; *Electronic Literature*, 11–12; *How We Think*, 117, 182, 190n5
Heidegger, Martin, 7–8, 15, 19, 179
Herbert, Zbigniew, 117–118
Here I am (Foer), 13, 16, 57, 77–80, 95–110; Eggers and, 132; Ferris and, 98, 100, 103, 108; Hamid and, 165, 170; Pynchon and, 61; Zadie Smith and, 162
Heschel, Abraham Joshua, 106
Holiday, Billie, 194n4
Hologram for the King, A (Eggers), 13, 114
How to Get Filthy Rich in Rising Asia (Hamid), 164, 194n14
How We Think (Hayles), 117, 182, 190n5
Huehls, Mitchum, 55
Humann, Heather Duerre, 74
human rights, 114, 133, 139, 144–145, 192n3
Hungerford, Amy, 193n16
Hutcheon, Linda, 190n3
Hutton, Margaret-Anne, 140
Hwang, Jung-Suk, 118
hybridity, 112, 147, 183–184; Burke on, 8, 14; Egan on, 64, 70–75; Hamid on, 149–150, 187; Haraway on, 77, 112, 184; Hayles on, 102; Shteyngart on, 20; Zadie Smith on, 147, 153. *See also* Bhabha, Homi K.
hypertext fiction, 11–12, 73

identity theft, 1, 57, 181
Ihde, Don, 7, 8, 101
Inherent Vice (Pynchon), 6
interactive fiction, 11, 171–172
intersectionality, 184, 195n2
"In the Ruins of the Future" (DeLillo), 113, 116, 125
intimacy, 9, 19; Pynchon on, 62; Roupenian on, 35; "virtual," 97, 166
Iraq, 6
"Is Google God?" (Friedman), 76, 77
"Is It O.K. to Be a Luddite?" (Pynchon), 47–49, 61, 190n3

Jackson, Melanie, 61, 63
Jackson, Shelley, 11–12
Jameson, Fredric, 10, 113, 180, 182–183
Jeffrey, David Lyle, 92

Johnston, John, 13–14
Joyce, Michael, 11–12
Judaism, 81; afterlife and, 96; Ferris on, 90–91; Foer on, 95–110, 192n13
Judd, Donald, 142, 143
Junod, Tom, 192n1

Kermode, Frank, 59
Kernan, Alvin, 189n2
Keyes, Ralph, 88, 103, 191n6
Khazan, Olga, 22
Kierkegaard, Søren, 8
King, Martin Luther, Jr., 7
Kingsolver, Barbara, 13
Krauss, Nicole, 102
Kruger, Anique, 148
Kubrick, Stanley, 11
Kundera, Milan, 27–28

Laist, Randy, 192n2, 192n4
Landrum, Ashley, 3
Lanier, Jaron, 8, 56–57
Lankshear, Colin, 181
Lechner, Frank J., 76, 152
LeClair, Tom, 192n2
Leps, Marie-Christine, 122
Life on the Screen (Turkle), 14, 96–97, 149
Lin, Tao, 13
literacies, 15, 18, 180–184; Griffey on, 190n5; Hayles on, 190n5; New London Group on, 31, 90, 183
Little Failure (Shteyngart), 23, 190n4
Loh, Lucienne, 193n4
Look at Me (Egan), 13, 49
Luce, Henry, 1, 51, 152, 184, 185
Luddites, 61, 78, 190n3
Luther, Martin, 85
Lynch, Michael Patrick, 5, 55

Mączyńska, Magdalena, 193n1
magical realism, 111; of Foer, 191n2; of García Márquez, 195n15; of Hamid, 164, 173, 175, 177, 187
Malewitz, Raymond, 24–25
Manjoo, Farhad, 180
Manning, Chelsea, 6
Mao II (DeLillo), 112
Mao Zedong, 120
Marcus, Sharon, 182
Marx, Karl, 120–121

Massive Open Online Courses (MOOCs), 3–4
Mather, Cotton, 88, 191n7
Matrix, The (films), 11
Mauchly, John, 189n5
McCann, Colum, 6
McChesney, Robert W., 9–10, 111
McClure, John A., 193n11
McLaren, Peter L., 181
"meatspace," 48, 56. *See also* cyberspace
"Meet the President!" (Zadie Smith), 13, 17–18, 98, 148, 151–163, 183
Melville, Herman, 34, 135
#MeToo movement, 6–7, 22
Milano, Alyssa, 6–7
Miller, Vincent, 12–13
Misa, Thomas J., 84–85
Moling, Martin, 72
"More Abandon (Or Whatever Happened to Joe Pope?)" (Ferris), 82
Morey, Peter, 194n5
Morgan, Katrina, 64
Moth Smoke (Hamid), 164
multiliteracies, 18, 31, 78, 90, 183, 184. *See also* literacies
Myanmar, 185

Nabokov, Vladimir, 27
Native Americans, 167
Natural-Born Cyborgs (Clark), 42, 69, 116, 136, 149, 183
Nealon, Jeffrey T., 10, 180, 182, 190n9
neoliberalism, 5; Shteyngart on, 23; terrorism and, 53–54, 65–66. *See also* globalization
New London Group, 31, 90, 183
9/11 attacks (2001), 2–3, 99; Chilean coup and, 53–54; DeLillo on, 112, 192n1; Egan on, 48, 65, 69–72; Pynchon on, 50–54, 56–64
Noble, David F., 3, 77, 82
"no child left behind," 55
nostalgia, 11; American, 64; for Cold War, 51–53; Egan on, 75; flattening effect of, 67; for future, 49; Hamid on, 169; Wajcman on, 48
NW (Zadie Smith), 148, 162, 193n1, 194n12

Oates, Joyce Carol, 22
Occupy Wall Street, 7
"O.K., Glass" (Shteyngart), 20, 24
Olster, Stacey, 147
On Beauty (Zadie Smith), 148, 193nn1–2

online dating, 19–20, 22, 35–37, 40, 134. *See also* social media
outsourcing, 4

Pannapacker, William, 3–4
Pariser, Eli, 186, 195n4
Park, Chung Won, 31
phonautograph, 71
Plato, 140
poetry, flash, 11
Point Omega (DeLillo), 13
pornography, 26, 38, 43
post-postmodernism, 10, 59, 113, 182, 190n9
"Power of History, The" (DeLillo), 129
PowerPoint software, 50, 72, 183, 193n1
Powers, Richard, 12
Prabhu, Anjali, 8
Presley, Elvis, 120
Pressed for Time (Wajcman), 9, 46–47, 112, 125
Pressman, Jessica, 14, 182
"Primer for the Punctuation of Heart Disease, A" (Foer), 79
print culture, 14, 102; Eggers on, 115; Ferris on, 84–85; Franzen on, 171, 189n2, 193n10
Proust, Marcel, 66
Pursell, Carroll, 1, 4, 98, 116
Pynchon, Thomas, 47–50, 57, 184; on "badasses," 48, 50, 61–62, 64, 75, 191n3; on digital technology, 115; engineering career of, 47; post-postmodernism and, 59
—works of: *Against the Day*, 55; *The Crying of Lot 49*, 47–48, 50, 54, 58, 63; "Entropy," 52; *Gravity's Rainbow*, 47; *Inherent Vice*, 6; "Is It O.K. to Be a Luddite?," 47–49, 61, 190n3. *See also Bleeding Edge*

racism, 32, 167–168, 170, 173, 184. *See also* xenophobia
reality television, 135–136
Reclaiming Conversation (Turkle), 19–20, 25, 84, 90, 97, 187
Reluctant Fundamentalist, The (Hamid), 164, 194n5
"Rewiring the Real" (Egan), 48, 70, 73
robots, 8, 11, 136, 166. *See also* cyborgs
Rohingya, 185
Rothko, Mark, 118, 142
Roupenian, Kristen, 20–23, 179; Egan and, 70; Foer and, 100

—works of: "The Boy in the Pool," 22; "Cat Person," 13, 15, 20–23, 34–45, 134, 169; "Death Wish," 22; "The Good Guy," 40; "Scarred," 40; "What It Felt Like," 22
Rushdie, Salman, 174, 195n15
Russell, Alison, 78
Russian Debutante's Handbook, The (Shteyngart), 20
Ruthven, Malise, 114–115

Salvation Army, 93
Sample, Ian, 3, 85–86
Santa Clara v. Southern Pacific Railroad (1886), 115
"Scarred" (Roupenian), 40
Schmidt, Eric, 56
Schradie, Jen, 189n5
Schreibman, Susan, 4
Schulz, Bruno, 79
science fiction, 10–18, 187, 190n10; Attebery on, 190n10; of Cold War, 10–12; Shteyngart on, 24
Sciolino, Martina, 118–119
Scopes Trial (1925), 76
Scott de Martinville, Édouard-Léon, 71
search engines, 3; Egan on, 73–74; Ferris on, 82, 86; Pynchon on, 47, 54–56, 73; Shteyngart on, 25–26; Zadie Smith on, 153, 162
Second Self, The (Turkle), 9, 164, 165
Seed, David, 10, 12
sexually explicit text messages ("sexts"), 99–100
Shakespeare, William, 106
Shelley, Mary, 12
Shirky, Clay, 189n3
Shteyngart, Gary, 20–21, 179, 190n1; Egan and, 70; on globalization, 23, 31
—works of: *Little Failure*, 23, 190n4; "O.K., Glass," 20, 24; *The Russian Debutante's Handbook*, 20. *See also Super Sad True Love Story*
Silence, The (DeLillo), 13, 112, 122
Silicon Alley, 55
Silicon Valley, 1, 5, 114–115, 150
slacktivism, 6–7, 184
slavery, 24, 134
Smith, Roberta, 142
Smith, Zadie, 148–150, 180; "Crazy They Call Me," 194n4; "Elegy for a Country's Seasons," 152, 194n13; "Generation Why?," 10, 73; "Meet the President!," 13, 17–18, 98, 148, 151–163, 183;

NW, 148, 162, 193n1, 194n12; *On Beauty*, 148, 193nn1–2; *White Teeth*, 147–148, 193n2
social media, 19–20, 22, 35–37, 40, 89, 134; Egan on, 74; Eggers on, 114, 134; Ferris on, 77–78; Foer on, 109; Friedman on, 19; Hamid on, 166; Zadie Smith on, 148, 163
Sprecher, Susan, 36
Starre, Alexander, 31
Steger, Manfred B., 112, 118, 146
Stephenson, Neal, 11
Stevens, Wallace, 49
Streisand, Barbara, 86
Sunstein, Cass R., 189n3
Super Sad True Love Story (Shteyngart), 13, 15, 20, 23–33, 180; print culture and, 85, 183; Slavic literature in, 52
surveillance, 17, 55–56; Egan on, 69; Eggers on, 132–134, 141; Hamid on, 166, 168

Talmud, 108
Tanakh, 103, 104, 107–108. *See also* Bible
"Technology Is Diminishing Us" (Foer), 95, 110
terrorism: capitalism and, 53–54, 65–66; digital culture and, 60–61; gender and, 62–63
Then We Came to the End (Ferris), 77, 78, 82
Tolstoy, Leo, 27
To Rise Again at a Decent Hour (Ferris), 13, 16, 57, 77–95; DeLillo and, 119, 126; Foer and, 98, 100, 103, 108; Hamid and, 168, 173; Zadie Smith and, 162
Tree of Codes (Foer), 79, 80, 110
Treisman, Deborah, 45
Trendel, Aristi, 121, 192n4
Trump, Donald, 146, 150, 184; fake news and, 181; as Twitter President, 195n1
Turkle, Sherry: *Alone Together*, 19, 97, 123, 150, 165, 166; *Life on the Screen*, 14, 96–97, 149; *Reclaiming Conversation*, 19–20, 25, 84, 90, 97, 187; *The Second Self*, 9, 164, 165
"twin paradox," 122, 192n7
Twitter. *See* social media

"Ulmism" (Ferris), 78, 87–88, 90–91
Underworld (DeLillo), 7, 12, 112, 124–125, 127, 192n1
Updike, John, 12
U.S. exceptionalism, 16, 75, 147, 184–185; Egan on, 69, 75; Pynchon on, 15, 47. *See also* American Century

Vanderwees, Chris, 98
video games, 51, 60, 99, 156–157
"virtual intimacy" (Turkle), 97
virtual reality, 8, 11, 101; DeLillo on, 117; Egan on, 69; Eggers on, 134, 140; Ferris on, 82, 95; Foer on, 96–97, 98, 170; Hamid on, 174; Zadie Smith on, 17–18, 149, 153, 155–162, 183
Visit from the Goon Squad, A (Egan), 13, 15–16, 47–50, 64–75, 175, 183; DeLillo and, 66, 73, 121; Pynchon and, 65, 66, 73, 74; Smith and, 73
Voice of Witness (organization), 114, 192n3
Vollen, Lola, 192n3
Vonnegut, Kurt, 11

Wajcman, Judy, 48; *Pressed for Time*, 9, 46–47, 112, 125
Wallace, David Foster, 34
Wanner, Adrian, 20
Warhol, Andy, 120
Weber, Max, 92, 191n8
Weinstein, Harvey, 22
What Is the What (Eggers), 114
"What It Felt Like" (Roupenian), 22
WhatsApp, 166
White Noise (DeLillo), 112
White Teeth (Zadie Smith), 147–148, 193n2
WikiLeaks, 6, 86
Wikipedia, 3, 87–88
Winthrop, John, 80, 152, 191n3, 194n9
Won, Esther, 190n1
Woodstock concert (1969), 65
World Is Flat, The (Friedman), 4, 18, 146, 185; Ferris and, 78; "Globalization 3.0" and, 1–3, 46, 111, 179
Wurth, Kiene Brillenburg, 79–80, 110

xenophobia, 146, 147, 161, 166; Ferris on, 83; Hamid on, 17–18, 150, 168, 173, 177, 184. *See also* racism

YouTube, 3, 5, 89, 184
Y2K scare, 1, 2, 51

Zeitoun (Eggers), 114
Zionism, 101
Zuckerberg, Mark, 148

www.ingramcontent.com/pod-product-compliance
Lightning Source LLC
Chambersburg PA
CBHW011742220426
43665CB00024B/2905